Introduction to Metaverse

Rajan Gupta · Saibal K. Pal

Introduction to Metaverse

Technology Landscape, Applications, and Challenges

Rajan Gupta
Artificial Intelligence and Innovation
(AI&I) Lab
Autonomous University
of Tamaulipas
Tamaulipas, Mexico

Saibal K. Pal
DRDO
Delhi, India

ISBN 978-981-99-7396-5   ISBN 978-981-99-7397-2   (eBook)
https://doi.org/10.1007/978-981-99-7397-2

Cover illustration: © Melisa Hasan

This Palgrave Macmillan imprint is published by the registered company Springer Nature Singapore Pte Ltd.
The registered company address is: 152 Beach Road, #21-01/04 Gateway East, Singapore 189721, Singapore

Paper in this product is recyclable.

Dedicated to

My "Gurumaa" *for holding my hand in Life,*
My "Parents" *for making me stand in Life,*
My "Nephews—Reyaansh & Atharva" *for making me strong in Life,*
My "Brother & His Wife" *for helping me progress in Life, and*
My "Wife" *for supporting and loving me unconditionally in Life!*

—*Dr. Rajan Gupta*

The memory of my "Parents"...

—*Dr. Saibal K. Pal*

PREFACE

Businesses are moving towards a future that is more automated and technologically equipped!

This book examines Metaverse technology, its definition, its characteristics, the factors affecting businesses, its adoption among businesses, its scope in the public/private sector, and its scope today and in the future. It also examines the metaverse's value creation potential and how the policymakers and leaders can plan and strategies to diligently make the most of this technology along with near-term actions.

The book begins with the introduction of the term metaverse, its elements, and how it has evolved till now. In-depth desk-based research has been conducted in this area with the help of government reports and organizational reports to extract suitable information. While introducing the metaverse, it was found that there is no single definition available to the concept rather it is evolving with every revolutionary discovery around this area. Some of the main elements (terms, use-cases, techniques, etc.) around the metaverse, that could be discovered during this research includes online shopping, digital humans, workplace, Natural Language Processing, gaming, social media, NFTs, and digital assets. It is examined that this shift is quite similar to the mobile era where older devices were replaced by newer ones. It is expected that by 2028, the vision of the metaverse will become more clear and easy to be managed by both individuals as well as organizations.

The metaverse is found to have the capability to impact everything from customer experience to employee engagement, product innovation as well as community building. The adoption of this technology is expected to accelerate in the next decade. This book examined seven layers of the metaverse which correspond to experience, discovery, creator economy, spatial computing, decentralization, human interface, and infrastructure. These components have become a strong building block of the metaverse.

Multiple factors are driving the metaverse towards growth. While this estimation, popular metaverse tools are emerging in this area and are already gaining popularity in the market. Decentraland has become a pioneer of metaverse technology which helps people in exploring their creative side through artworks and other virtual experiences. Sandbox is another Ethereum blockchain, an NFT gaming metaverse which has become popular due to its conglomeration of its three integrated products. These are VoxEdit, marketplace, and game maker enjoying 39,000 daily users. Illuvium, Roblox, and Cryptovexels are other metaverse platforms that are gaining popularity which has created an altogether different virtual world providing an immersive experience to the users.

Popular companies such as Apple, Accenture, Meta, Disney, and Microsoft are also leveraging the metaverse platforms. Facebook or Meta was among the first set of companies which brought the concept of Metaverse to the notice of common users in 2021, however, the concept was released by Author Neal Stephenson in his 1992 science fiction novel about life in virtual reality, "Snow Crash." The focus is on connecting people seamlessly, finding suitable communities, and growing businesses. Tech giant Apple has also opted for a unique approach for the metaverse but it is necessary to say that its application is still in the developmental phase. Microsoft has its own distinctive vision for the metaverse. Its "Mesh" introduction for Teams has been serving as a key pathway to the persistent digital realm. It is expecting to use metaverse in the areas of entertainment, business, training, and education. Entertainment companies like Disney are also ripping the concept by offering immersive and experiential online storytelling as a part of its metaverse business strategy. Putting all these aspects together, it is important to note that the technology is still developing but large tech companies are leaving no stone unturned to make the most of what is available in the course of the metaverse.

The potential impact of the metaverse varies from one industry to the other. Gaming was the first industry that early on detected the potential of the metaverse and began its adoption with the help of 3D avatars and gameplay. Building a connected universe of virtual reality is helping in providing realistic touch to the users. This growth is constantly contributing to the surge of online game adoption with a CAGR of 47.6%. The education industry is also inclining towards providing immersive learning experiences to its learners through concept visualization. Healthcare and Pharma industry is adopting the metaverse in the hospitals and pharmaceutical companies joining in to improve patient care. Pharmaceutical industries are experiencing a faster supply chain and reduced training costs in comparison to traditional methodologies. It has been helping companies in developing lasting competitive advantages.

Government is on an altogether different journey of the metaverse. The adoption begins by becoming a modern digital government by using open data across all channels. It has been making the communication process more clear and lucid which further assists in the decision-making process. Although metaverse poses challenges for the government as the existing provisions demonstrate a lack of online security posing risk to the users. Other top barriers identified are insufficient technical skills, lack of collaborative and shared culture, lack of understanding, and too many competing priorities. The major question that arises in this area is how will the government handle misinformation and disinformation. Also, how will the government ensure regulatory compliance as well as tax reporting? It is undoubtedly a complex regulatory that could require collaboration with the private sector for exploring regulatory approaches.

As a part of the way forward, the metaverse is the next version of the internet and privacy can become a major cause of concern in this area. If data rested in the wrong hands, it can be a huge loss to goodwill and the wealth of people as well as the nation. Hence, it requires preparedness from a security and privacy point of view to avoid the technology from becoming a threat to humans. Ethical aspects are also crucial in the metaverse. For the metaverse to become truly inclusive, it becomes important to work on the existing inequalities and develop a digital environment suitable for everyone's needs. Though it is expected that the metaverse will revolutionize in the upcoming years, it needs constant vigilance and global actions to be taken in a concentrated manner. As far as private organizations are concerned, ROI considerations will be really important after identifying relevant use-cases, before investing in the Metaverse Technology.

Delhi, India   Rajan Gupta
Delhi, India   Saibal K. Pal

ACKNOWLEDGEMENTS

The authors of this book would like to gratefully and sincerely thank all the people who have supported them during the journey of writing this book, only some of whom it is possible to mention here.

Primarily, the authors would like to thank their Ph.D. Supervisor—Prof. Sunil Kumar Muttoo, for his valuable guidance and research directions in the field of Computer Science and Data Science. Then the authors would like to thank current and former faculty members of the Department of Computer Science, University of Delhi—Prof. Naveen Kumar, Prof. Vasudha Bhatnagar, Prof. Punam Bedi, Prof. Neelima Gupta, Mr. P. K. Hazra, and Ms. Vidya Kulkarni. Also, the author (Dr. Gupta) would like to thank faculty members from the Center of Information Technologies and Applied Mathematics, University of Nova Gorica, Slovenia, led by Prof. Tanja Urbancic, Prof. Irina, Prof. Nada, and Ms. Tea for their valuable support. And a special mention to Prof. Devendra Potnis from UTK, USA for his extremely helpful suggestions around developing research areas around Algorithmic Government. They all helped provide infrastructure and resources related to doctoral and post-doctoral research work, which was in Technology, Data Science, and Public Information Systems. The doctoral as well as post-doctoral research work helped in forming the basis for this book.

The author (Dr. Gupta) would like to thank Dr. Fernando Ortiz-Rodriguez and Dr. Sanju Tiwari from AI&I Lab at UAT, Mexico, for providing valuable support towards writing this book. The author (Dr.

Gupta) would also like to thank all the members of Analyttica Datalab, especially Research & Analytics Division, as well as all the members of TCF Consultancy. The author (Dr. Pal) would like to thank the Defense Research & Development Organization (DRDO) authorities, Government of India, for providing valuable support towards the research work carried out for this book.

Finally, this work would not have been possible without the invaluable support from the reviewers, editors, and the entire publishing team of Palgrave Macmillan, Springer, esp. Ms. Aparajita Singh, Ms. Aurelia Heumader, and Mr. Ananda Kumar Mariappan. This book also recognizes incredible support from the book's endorsers, and the authors' guru, mentors, family, and friends. So the authors would like to thank them all from the bottom of their hearts.

CONTENTS

1 **Concept of Metaverse**   1
   1.1   *Defining Metaverse*   1
   1.2   *Elements of Metaverse*   4
         1.2.1   Online Shopping   4
         1.2.2   Digital Humans   4
         1.2.3   Workplace   5
         1.2.4   Natural Language Processing (NLP)   5
         1.2.5   NFTs and Digital Assets   5
         1.2.6   Gaming   6
         *1.2.7   Social Media, Concerts, and Entertainment
                  Events*   6
   1.3   *Evolution of Metaverse*   7
         1.3.1   Phase 1: Emerging Metaverse   7
         1.3.2   Phase 2: Advanced Metaverse   9
         1.3.3   Phase 3: Mature Metaverse   11
   1.4   *Metaverse as a Game Changer*   11
   1.5   *Why Metaverse Continuum?*   13
   1.6   *Economic Implications of Metaverse Technology*   15
   References   18

2 **Metaverse in the Technological World**   23
   2.1   *Metaverse vs Web 2.0 vs Web 3.0*   24
   2.2   *Blockchain Technology*   27
   2.3   *Digital Twin*   30

2.4	Metaverse vs Multiverse	33
2.5	Artificial Intelligence (AI)	35
2.6	Internet of Everything (IoE)	37
2.7	Hyperautomation	39
	References	41

3 Seven Layers of Metaverse — 45

3.1	Introduction to Seven Layers of Metaverse	45
3.2	Layer 1—Experience	47
3.3	Layer 2—Discovery	48
3.4	Layer 3—Creator Economy	50
3.5	Layer 4—Spatial Computing	53
3.6	Layer 5—Decentralization	55
3.7	Layer 6—Human Interface	58
3.8	Layer 7—Infrastructure	59
	References	61

4 Metaverse Platforms and Use-Cases — 67

4.1	Major Platforms and Tools	67
	4.1.1 Decentraland	68
	4.1.2 Sandbox	68
	4.1.3 Illuvium	69
	4.1.4 Axie Infinity	70
	4.1.5 Cryptovoxels	70
	4.1.6 Roblox	71
	4.1.7 Metahero	72
4.2	Global Outlook	73
	4.2.1 Apple	73
	4.2.2 Facebook/Meta	74
	4.2.3 Microsoft	75
	4.2.4 Disney	76
	4.2.5 Accenture	77
4.3	Industrial Use-Cases and Business Implications	78
	4.3.1 Gaming	78
	4.3.2 Education and Learning	78
	4.3.3 Healthcare and Pharmaceuticals	80
	4.3.4 Travel and Tourism	81
	4.3.5 Banking and Finance	81

4.3.6   Human Resource Management   82
4.3.7   Social Media and Entertainment   83
References   84

5   Metaverse for Public Sector   91
5.1   Evolution and Digital Maturity of Government   91
       5.1.1   Level 1: Initial (E-Government)   92
       5.1.2   Level 2: Developing (Open)   92
       5.1.3   Level 3: Defined (Data-Centric)   92
       5.1.4   Level 4: Managed (Fully Digital)   93
       5.1.5   Level 5: Optimizing (Smart)   93
5.2   Challenges for Government   93
5.3   Metaverse as a Solution: Role and Different Models   95
       5.3.1   Broadcasting Model   96
       5.3.2   Critical Flow Model   96
       5.3.3   Interactive Services Model   96
       5.3.4   Mobilization and Lobbying Model   97
       5.3.5   Comparative Analysis Model   97
5.4   Potential Use-Cases of Metaverse for Government   98
       5.4.1   Aerospace and Defence   98
       5.4.2   Art and Culture   100
       5.4.3   Trade and Economy   101
       5.4.4   Employment   102
       5.4.5   Public Health and Safety   104
       5.4.6   Public Entertainment   106
       5.4.7   Knowledge Management   106
5.5   Implications of Metaverse Adoption for Government   107
5.6   Public Policy for Metaverse   108
References   109

6   Way Forward For Metaverse Adoption   115
6.1   Business Framework for Metaverse Adoption   115
6.2   Challenges of Metaverse Technology   119
       6.2.1   Privacy Issues   119
       6.2.2   Fairness   120
       6.2.3   Cyberbullying   120
       6.2.4   Social Issues   121
       6.2.5   Accountability   121
       6.2.6   Safety   122
6.3   Preparedness for Metaverse Technology   122

6.3.1	*Human Resource Preparedness*	122
6.3.2	*Security and Privacy*	123
6.3.3	*Threat to Humans*	124
6.4	*Potential Misuse of Metaverse Technology*	125
6.5	*Ethical Considerations for Metaverse*	126
6.6	*Business Implications*	127
6.6.1	*For Consumers*	127
6.6.2	*For Enterprises*	128
6.6.3	*Financial Implications*	130
6.7	*Preparing for the Future: What Can Be Done Today*	132
6.8	*Concluding Remarks*	135
	References	137

References   143

Index   165

About the Authors

Dr. Rajan Gupta is a Research and Analytics Professional with 14+ years of combined experience in AI/ML Product & Services Delivery, Analytical Research, Consulting, Training, and Teaching in the field of Data Science and Computer Science. He has authored 100+ publications including 7 books and multiple research papers in the area of Public Information Systems, Artificial Intelligence (AI), Machine Learning (ML), Data Science, Information Technology, and Management. He has received the prestigious **"AI Changemaker Leader"** under 3AI ACME Awards at BEYOND 2023; **"AI Makers 100"** which is the Top 100 Most Influential AI & Analytics Leaders Award at 3AI GCC-X Summit 2023; and **"40 Under 40 Data Scientists"** award for 2022 by Analytics India Magazine at MLDS 2022.

He has recently headed the Research & Analytics (R&A) Division of Analyttica Datalab Inc (as Vice President—Research & Analytics), which is a contextual Data Science & Machine Learning product company in experiental AI/ML & Analytics EdTech space at the global level, and is currently a visiting researcher at AI&I Lab, UAT Mexico. He has exposure to working with Fortune 500 global clients in Consulting, Telecom, Automotive, Healthcare, Retail, and Manufacturing domains, and global research labs for the United Nations and European Commission.

He has done his Ph.D. in Information Systems & Analytics from the Department of Computer Science, University of Delhi and Postdoc in Data Science and Data Modelling from Center of Information Technologies and Applied Mathematics, University of Nova Gorica, Slovenia, Europe. He is a triple post-graduate and completed his Master in Computer Application (MCA) from the University of Delhi; Post Graduate Program in Management (PGPM) from IMT Ghaziabad (CDL); and Executive Program in Business Intelligence and Analytics (EPBABI) from IIM-Ranchi. He is UGC NET-JRF qualified and holds a certificate in Consulting from Consultancy Development Centre (CDC), DSIR, Ministry of Science and Technology, Government of India. He is one of the few Certified Analytics Professional (CAP-INFORMS) around the world and is serving as CAP Ambassador in Asia Region. He is the first non-US member of the prestigious Analytics Certification Board (ACB) of INFORMS, USA. He has also been accredited with "Graduate Statistician" by the American Statistical Association (ASA).

He has worked on various analytical consulting assignments at national and international levels on problems related to the areas of Digital Government, Healthcare, Education, Retail, and Insurance. He has delivered lectures at the University of Delhi and IMT—Ghaziabad in Computer Science, Data Science, Information Security, and Management. He has also conducted several 1:1 live mentoring and training sessions related to upskilling of analytical career, preparations for analytical certifications, and technical knowledge on Analytics Project Lifecycle for various data science professionals from reputed organizations like Verizon, Nokia, Cyient, Tech Mahindra, TCS, Ericsson, and Anthem.

His area of interest includes Artificial General Intelligence (AGI), Generative AI (GAI), EdgeAI, Metaverse, Algorithmic Government, Hyperautomation, Network Science, Data Science, E-Governance, Public Information Systems, and Information Security. He has contributed to the E-Governance Development Index report by the United Nations (EGDI-2020). He is a member of the reviewer panel of multiple international journals and conferences. He has also delivered a talk as a panellist on Data Science Application for E-Governance on an international forum sponsored by International Data Engineering and Science Association (IDEAS), USA, and conducted a Global Workshop on "Inclusion of Marginalized Communities" through Electronic Governance and Analytics at ICEGOV-2020 hosted by United Nations University, among many Corporate events, Panel Discussion, Enterprise Webinars, Faculty

Development Programs, News Channel Debates and Research events, as an AI/ML & Data Analytics expert.
LinkedIn Profile: https://www.linkedin.com/in/rajan-gupta-cap/

Dr. Saibal K. Pal is a Senior Scientist with Scientific Analysis Group (SAG) Lab, Defense Research & Development Organization, Government of India, for many years and has been awarded "Scientist of the Year" by the Government of India. He received his Ph.D. in Computer Science from the University of Delhi and is an Invited Faculty and Research Guide at several national institutions. His areas of interest are Information & Network Security, Computational Intelligence, Information Systems, and Electronic Governance. He has more than 250 publications in books, journals, and international conference proceedings. He has contributed to a number of significant projects and international collaborations and is a member of national advisory committees.

ABBREVIATIONS

3D	Three-Dimensional
AI	Artificial Intelligence
AIFUCT	All India Federation of University and College Teachers
APAC	Asia Pacific
API	Application Programming Interface
AR	Augmented Reality
CAGR	Compound Annual Growth Rate
CCDH	Centre for Countering Digital Hate
CRM	Customer Relationship Management
DAO	Data Access Objects
dApps	Decentralized Applications
DeFi	Decentralized Finance
DID	Decentralized Identifiers
DXP	Digital Experience Platforms
ETH	Ether
FICCI	Federation of Indian Chamber of Commerce and Industry
GDP	Gross Domestic Product
GDPR	General Data Protection Regulations
HCI	Human–Computer Interaction
HDM	Head–Mounted Displays
ICT	Information and Communication Technology
IMA	Indian Medical Association
IMR	Infant Mortality Rate
IoE	Internet of Everything
IoMD	Internet of Medical Devices
IoT	Internet of Things

IPR	Intellectual Property Rights
LATAM	Latin America
M2M	Machine to Machine
MEMS	Microelectromechanical systems
MENAT	Middle East, North Africa and Turkey
ML	Machine Learning
MMR	Maternal Mortality Rate
NFT	Non-Fungible Tokens
NLP	Natural Language Processing
NPC	Non-Playing Characters
OpenGL	Open Graphics Library
OS	Operating System
P2M	People to Machine
PC	Personal Computers
Q-Commerce	Quick Commerce
RPA	Robotic Process Automation
RTO	Regional Transport Office
SDK	Software Development Kit
SLAM	Simultaneous Localization And Mapping
SSA	Sub-Saharan Africa
UN	United Nations
VPN	Virtual Private Network
VR	Virtual Reality
XR	Extended Reality

LIST OF FIGURES

Fig. 1.1   The extent of Metaverse (*Source* Accenture [2022b])   3
Fig. 1.2   Evolution of Metaverse by years (*Source* Forbes [2022a,
            2022b])   8
Fig. 1.3   Evolution spectrum of Metaverse (*Source* Gartner [2022c])   8
Fig. 2.1   Blockchain plans by enterprises (*Source* IBM, 2022)   28
Fig. 2.2   Industry wise distribution of Global Digital Twin Market
            in 2020 (*Source* Statista, 2022)   31
Fig. 2.3   Global AI software market revenue (*Source* World
            Economic Forum, 2018)   35
Fig. 2.4   AI adoption around the world (*Source* Watson, 2022)   36
Fig. 2.5   Elements of IoE (*Source* Bandara, 2016)   38
Fig. 3.1   Seven layers of the metaverse (*Source* Far and Rad [2022]) 46
Fig. 3.2   Survey of online US adults on online activity (*Source* Pew
            Research Center [2022a])   48
Fig. 3.3   Evolution of creator economy (*Source* Innovius Research
            [2022])   51
Fig. 3.4   How does Blockchain work? (*Source* PwC [2022])   56
Fig. 3.5   Metaverse market map (*Source* Market place Fairness
            [2021])   61
Fig. 4.1   Challenges to be addressed in Gaming Applications
            of Metaverse (*Source* Appinventiv [2022])   79
Fig. 6.1   Key pointers for business leaders to consider while adopting
            Metaverse   116
Fig. 6.2   Hype cycle for emerging technologies by Gartner (*Source*
            Gartner [2022e])   133

LIST OF TABLES

Table 1.1   Technological characteristics of Metaverse   12
Table 1.2   Contribution to GDP from Metaverse in 10 years   16
Table 2.1   Web 2.0 versus Web 3.0 approach to Metaverse   25
Table 2.2   Blockchain-based virtual economy segments   29
Table 2.3   Difference between Metaverse and Multiverse   34
Table 4.1   Metahero's 5 year revenue projections as per paper
             released in 2022   73
Table 4.2   Microsoft vs. Meta comparison towards the metaverse
             technology   76

CHAPTER 1

Concept of Metaverse

Abstract Metaverse, a hypothetical internet iteration, is a concept that is new to the world. This chapter begins with defining the concept in detail followed by the requisite elements of the Metaverse. Considering the constant evolution in this area, it is important to understand the current state of development and innovation through the three main phases, viz., the emerging phase, the advanced Metaverse, and the mature Metaverse. Further, the Metaverse has become a game changer, especially after Covid-19 and has become a huge support to the economies. Its economic implications are crucial and need emphasis with a detailed assessment.

Keywords Metaverse · Online shopping · Digital humans · NFTs · Natural Language Processing · Gaming · Social media · Metaverse Continuum

1.1 Defining Metaverse

The term "Metaverse" is a combination of the word "Meta," meaning beyond, and "universe." It refers to the idea of a next version of the internet. But does this definition reflect what many in business and technology words refer to when they use the word Metaverse?

1

As per Gartner (2022a, 2022b, 2022c, 2022d, 2022e, 2022f), the term "Metaverse" refers to a collective virtual space, which is brought about by the merging of physical reality and digital reality. This virtual space is a continuous environment that offers a more immersive experience. When fully developed, it will act as a platform that transcends the limitations of specific devices and paves the way for further advancements.

The concept of the Metaverse encompasses the integration of digital and physical existences, resulting in a virtually cohesive community. This virtual world allows us to carry out a variety of activities such as work, leisure, socializing, and transactions. It is at the early stage of evolution and there is no single definition that can encompass all the aspects of Metaverse. Themes around the basics of the Metaverse are still emerging and adding to the existing definition of the Metaverse (Nevelsteen, 2018). Here, the main point is, there is no singular virtual world, but it is the conglomeration of many worlds which are shaping and enabling people to extend and deepen their social interactions digitally. This is performed by adding a 3D layer of the web that helps in creating a more natural and authentic experience.

To help in getting more sense of how complex as well as vague the term "Metaverse" is, this exercise can be performed. By substituting the word "Metaverse" with "Cyberspace" in a sentence, the meaning remains largely unchanged in 90% of cases. This is due to the fact that the term refers to a general, often speculative, shift in our relationship with technology, rather than a specific type of technology. It is highly likely that the term will become outdated as new technological terminology arises to take its place.

Broadly speaking, technology companies often refer to Virtual Reality (VR), when they talk about the Metaverse. It can be characterized by the persistent existence of virtual worlds that continue to exist even if one is not playing. These companies might also be referring to Augmented Reality (AR) which focuses on combining the aspects of the physical and digital worlds. However, it does not require that these spaces are exclusively accessed by AR or VR.

As per Gartner, 25% of people will be spending minimum one hour a day in the Metaverse for shopping, work, social media, entertainment, and education (Gartner, 2022b). The Metaverse is often referred to as the next iteration of the internet, which started with isolated bulletin boards and standalone online locations. Over time, these locations evolved into sites within a shared virtual space, much like how the Metaverse is expected to grow and develop (Fig. 1.1).

The future of the Metaverse is interoperable and seamless which is very different from what is existing today. In the current scenario, the digital world is not compatible. For instance, a user requires individual accounts to access varied social media applications such as TikTok and Twitter and an individual account for accessing gaming consoles such as PlayStation or Xbox (Joy et al., 2022). However, the Metaverse will empower users to seamlessly consume digital goods and services, making it a single sign-on kind of a smooth experience.

Andrew Chow from *Time Magazine* underpins this vision and wrote, "Rather than having separate Twitter and Facebook accounts in which everything posted by the individual is owned by those corporations, Metaverse will enable to own their digital personhood of all the ideas and digital belongings irrespective of where goes" (Time, 2021). As an example, a person can effortlessly buy a digital item of clothing or accessory from one platform and then access it on another platform, rather than being limited to using the digital item only on the platform where it was originally purchased (YouTube, 2021).

Several early elements of the Metaverse can already be observed, although in fragmented form. These components are already producing

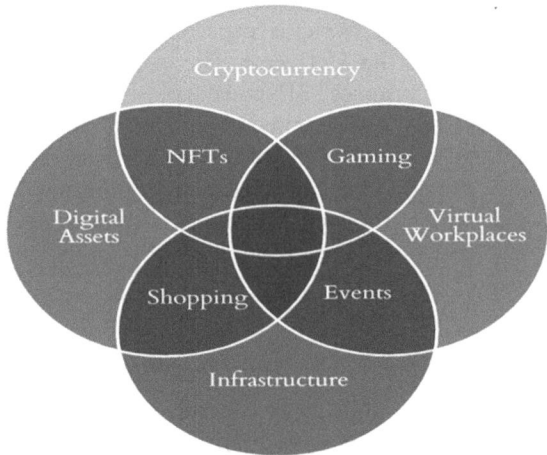

Fig. 1.1 The extent of Metaverse (*Source* Accenture [2022b])

commercial, social, and business effects. As users continue to adopt these technologies, their ability to bring about significant change in society will only increase. In future, the Metaverse is anticipated to evolve into a much more extensive entity, influencing the creation of new job opportunities, impacting conventional businesses and affecting consumers in innovative ways that are beyond our current imagination.

1.2   ELEMENTS OF METAVERSE

Each element plays an important role and overlaps most of the time in one or the other way. These elements are Digital Currency, Marketplace/Digital Commerce, Device Independence, Non-Fungible Tokens (NFTs), Online Shopping, Infrastructure, Gaming, Digital Assets, Workplace, Digital Human, Natural Language Processing and Concerts, Social Media, and Entertainment Events. The current set of elements forms the universe of Metaverse in terms of concepts, applications, and techniques hovering around it.

1.2.1   Online Shopping

According to Zhao et al. (2022), it goes without saying that the pandemic has changed various aspects of life, bringing people closer to the digital set-up and increasing their time spent online. The possibility of shopping online in the virtual world is expected to be one of the most exciting opportunities in the Metaverse. The overall process will become more experiential. For instance, 3D visualization will help in unlocking a whole new world of global commerce and retail trading. There is a virtual certainty that, if not all, most of the retailers will be taking up this impending transformation in online shopping (Darbinyan, 2022). A new study reveals that 70% of the Gen Z consumers (people born in twenty-first century) who visit virtual stores and then went on to make a purchase (Lim, 2022).

1.2.2   Digital Humans

According to Ahn et al. (2022), the Metaverse will also bring AI and, to some extent, AI to the Metaverse. It has the potential to increase productivity if opted correctly. However, it can also lead to service failure and cause problems. As companies grow, it is expected that there will be substantial growth in the corporation of AI bots in business practices which can further improve customer satisfaction as well. Generation of own AI avatars with deep customization in the likeness will endure reaching better digital opportunities.

1.2.3 Workplace

Some believe that the Metaverse is the future of remote work. With the advancement in technology, it can become easier to experience virtual reality than was once possible only in science fiction. It has led to the development of immersive workplaces allowing employees to perform work from home from anywhere in the world (Hawkins, 2022). Being an advanced form of remote working, a virtual workplace in the Metaverse will be able to provide the same experience and social interaction that employees get while standing around the coffee machines, during lunch breaks, and attending happy hours. It is expected that it will be a game changer in the workplace.

1.2.4 Natural Language Processing (NLP)

As per Goldberg (2016), the process of communicating with AI will require NLP which further requires a semantic web. Artificial code must be able to understand natural words and intonations. NLP will help take the interactions to a whole new level by making it possible to generate audio responses complete with voice modulations and linguistic nuances. It can evenly translate responses in multiple languages automatically to reach a wider audience. This is why Metaverse companies such as "Meta Platform" are involved in launching NLP aids with the developers (XR Today, 2022a, 2022b).

1.2.5 NFTs and Digital Assets

Non-Fungible Tokens (NFTs) fits perfectly into the Metaverse, which acts as authenticators and infallible provers of ownership made possible through blockchain technology. It takes cryptocurrencies such as Ethereum and Bitcoin by proving the ownership of unique objects. Some of the classic examples already available in the digital world are real estate and art.

As per ReportLinker.com, it is expected that the size of the global

NFT market will grow by $147.24 billion during 2022–2026 with an increasing CAGR of 35.27% during the forecasted period (ReportLinker, 2022). NFTs are easily divisible to custom specifications which enable fractional ownership of expensive items such as elaborate mansions and centuries-old paintings. Apart from collecting, it also opens up a whole new world of investing as well.

1.2.6   Gaming

It is the concept of playing games in the Metaverse. It is the permanent and solid pillar of the upcoming Metaverse. In such cases, players use virtual materials for upgrading their avatars or avatar "homes" while meeting with other players freely. As per the research, the gaming industry is the centre of Metaverse today. Having the first-mover advantage in this virtual world, gaming companies have already built the prototype of Metaverse and established their position as an early adapter in the industry (Nevelsteen, 2018).

As per the report published by EY, 97% believe that the gaming industry is the centre of Metaverse today (EY, 2022). For instance, widely played games, such as Fortnite and Minecraft as well as the popular Roblox game platform, have been able to incorporate many aspects of a Metaverse, including virtual worlds where players meet each other and use other social media features such as in-game chats. Various gaming companies are also providing in-game assets and in-game payment systems such as props, skin, clothes, vehicles, and weapons that travel with people across platforms (Consoles, PCs, Mobile devices, etc.).

1.2.7   Social Media, Concerts, and Entertainment Events

In the spirit of gaming, it is expected that most of the time, Metaverse might include personal experiences. Social media will become a hyper-personalized form, particularly with VR. Perhaps above all, Metaverse will be adopted by businesses as an extended ability of advertising products and services (Forbes, 2022a, 2022b). Just like the new form of advertising emerged through Web 2.0 social media, Web 3.0 will help in bringing for building hype and excitement around brands.

1.3 Evolution of Metaverse

The internet is undergoing a major transformation and the Metaverse is at the forefront of this change. Since the 1990s, the Metaverse has experienced significant growth. When we look back over the past few decades, we can see the progression from the early 1990s' Internet of Data to the 2000s' Internet of People, and the 2010s' Internet of Things (Accenture, 2022a, 2022b).

According to Vidal-Tomás (2022) it is expected that the next evolution will incorporate the internet of place and the internet of ownership along with Metaverse and Web 3.0. Back in the 2010s, the hype was all about how the adoption of Web 2.0 has emerged as a medium of mass participation. Then, in 2015, the hype was all about how technology has been amplifying people's lives. Today it has been moving towards more persistent shared experiences.

Currently, Metaverse seems everywhere yet nowhere, simultaneously persuading itself as the next great innovation and criticized as overhyped and over-promised at the same time. This transition is somewhat similar to the previous shifts in technology, such as the mobile era or industrial revolution, where new technology was introduced by the leaders displacing the obsolete ones. It is believed that the Metaverse will represent a significant advancement in internet development and function as a collective and shared space that arises from the merging of physical and long-lasting digital content and experiences (Fig. 1.2).

The Metaverse will evolve in three overlapping phases. These are emerging, advanced, and mature (Gadekallu et al., 2022). The below sections define what technology leaders need to know about each phase and present distinctive technology product influences or markets (Fig. 1.3).

1.3.1 Phase 1: Emerging Metaverse

Metaverse is currently in its emerging phase. It consists of the products and services that are commercially available such as social networks, cryptocurrencies, online games, NFTs, and e-commerce. Based on the requisite applications, aforesaid technologies can help in satisfying either one or more than one characteristic of the Metaverse such as decentralized, collaborative, persistent, and interoperability. However, it is crucial to note that the emerging Metaverse technology is still not complete

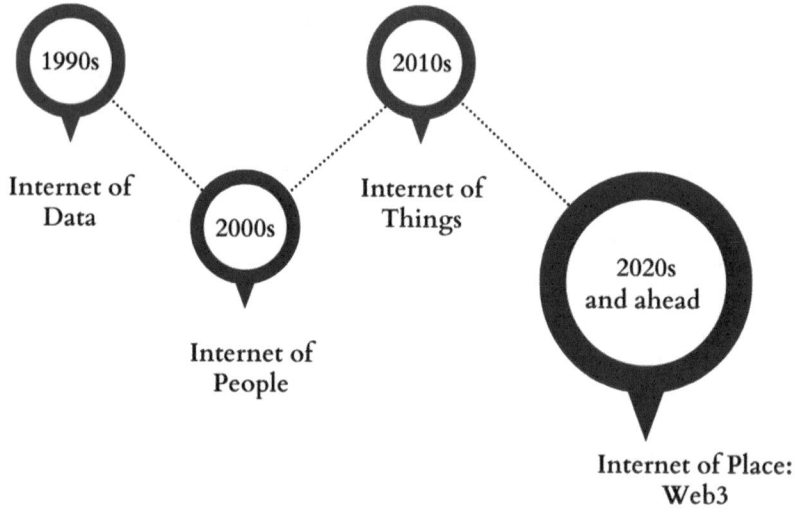

Fig. 1.2   Evolution of Metaverse by years (*Source* Forbes [2022a, 2022b])

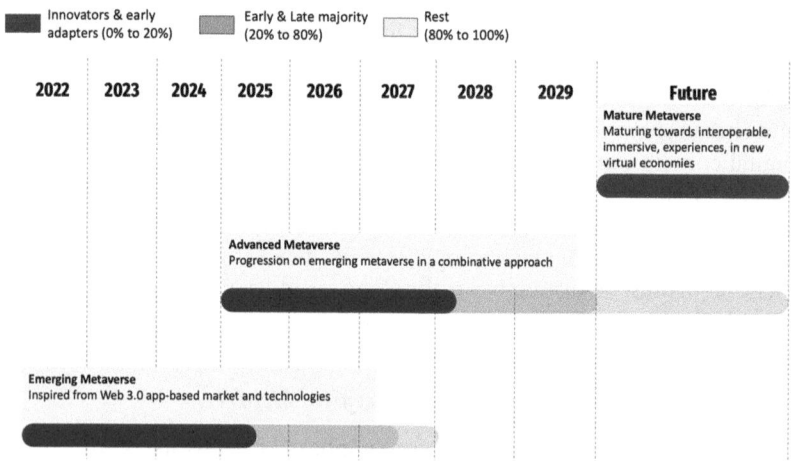

Fig. 1.3   Evolution spectrum of Metaverse (*Source* Gartner [2022c])

(Gadekallu et al., 2022). For instance, interoperability is still missing among different products and services that are marked as the Metaverse. The condition is not existing because of any deficiency in the offering but rather because there are no standard instructions or set-up of inter-operation. Therefore, all the existing applications and technologies that are called "Metaverse" today, these days are actually Metaverse-inspired solutions, re-metaverse, or mini-verse.

According to Wang (2022), to achieve widespread adoption of the Metaverse, any potential solution must provide the same capabilities as existing technologies in a user-friendly and accessible way. For example, currently augmented and virtual reality immersive experiences give us a glimpse of what the Metaverse could look like in the future, but they are still considered to be in a pre-metaverse state as users cannot transfer content between applications or move between applications themselves.

It is expected that the first phase of Metaverse will last through 2024. One of the major challenges that technology leaders will face during this phase will be creating a profitable and sustainable business model or justifying the ROIs based on use-cases of pre-metaverse such as VR and AR experiences. In the initial era of the Metaverse, the technology leaders must emphasize building and exploring pioneers in the Metaverse. Each key characteristic reflects the opportunity as well as the challenge in defining the product strategy of the Metaverse. Currently, high-value use-cases of the Metaverse, such as gaming, AR and VR experiences, and navigation and gaming apps, are some of the promising forerunners.

1.3.2 Phase 2: Advanced Metaverse

The cutting-edge Metaverse will be marked by the convergence of technology seen in its initial stages. It is expected that this phase will occur between 2024 and 2027 (BAE, 2021). This convergence will also focus on inspiring the building of newer technologies that will further enable mature Metaverse solutions, such as the method of linking digital and physical space in a navigable way.

One such example is Spatial computing technologies that are expected to be featured prominently in this phase. This technology allows for linking digital, long-lasting content to physical content, such as adding digital colours to Greek and Roman statues or determining the location and orientation of a digital object within a physical space, such as making sure a digital sign faces the right direction on a street.

Technology that occupies the information layer will also be developed during this phase. It includes innovations that focus on capturing, creating, and integrating digital content that overlays the physical world, such as technology mapping and sensing, places, people, processes, and things. It will also include graph technology establishing relationships and processes between different elements. Various other technologies also need to mature to support the advanced stage of the Metaverse which includes,

- Sensor fusion, sensor technologies, and data integration enabling technologies. This will enable systems in providing a higher context to make indexed and geo-posed content useful (Majerová & Pera, 2022).
- Products offering packaged business capabilities, such as Application Programming Interface (API), that focus on providing product and platform capabilities for other organizations which can be used as a base to build upon. It will also initiate composable Metaverse offerings to move away from the app-based, siloed approach to "super apps."
- Multimodal user interfaces streamline the shift towards device-agnostic experiences while providing innovative and easy-to-use methods for engaging with the merging physical and digital environment.
- Edge computing and EdgeAI are necessary for implementing sensing technologies like computer vision, interaction methods like NLP, and rendering high-volume, high-quality content.

Since, in phase 2, it is expected that the Metaverse will gradually mature, it is also important for the technology leaders to evolve their product portfolios at the same time that can further support Metaverse experiences (Majerová & Pera, 2022). For instance, by functioning with the standard and protocol groups for defining and tracking interoperability. Since no one company will build a Metaverse, it is important for the providers to create or join an ecosystem of service delivery partners.

1.3.3   Phase 3: Mature Metaverse

According to López-Belmonte et al., (2022), the matured phase of Metaverse will notice that most of the applications have facilities that can enable multi-sourced and collaborative experiences. Moreover, interoperable content across digital experiences will indicate the arrival of the matured phase. For example, imagine a digital note that is linked to a physical object like a traffic light, which can be easily accessed and modified by technicians. Additionally, citizens would also be able to access additional information about this object and report issues or complaints. Since the content is anchored digitally, all the citizens can have access to the information about when the outrage was reported instead of redundant efforts.

It is expected that spatially oriented, indexed, and persistent content could be adopted by some early movers by the year 2028, but the overall characteristics of the market of the matured Metaverse will be found prevalent from 2030 (Ooi et al., 2022). From the year 2028, the potential and vision for Metaverse will become easy and clear for managing both organizational and individual users. It will mainly be developed on inspirational use-cases and applications discovered during Phase 2, reinforced by matured other technologies, such as digital currencies, computer vision, immersive technology, 5G technology, and immersive technology. As such, functionalities and aspects of the systems required for making the matured Metaverse possible will be understood by opening appropriate opportunities in the infrastructure layer. Also, the vendors will compete for creating the backbone of a transformational and potentially ubiquitous system (Table 1.1).

1.4   METAVERSE AS A GAME CHANGER

The Metaverse concept is not new, rather it has a linear progression in many ways. Online, role-playing worlds and multiplayer models have been around for more than 20 years, and players spend approximately 20 hours of average time per week in this world. Modern equipment such as World of Warcraft, Minecraft, and Fortnite enjoy having numerous users as well as huge supporting economies.

The world has reached an inflexion point where there is no single day without a company or celebrity announcing their presence in the virtual space (Gartner, 2022c). Convergence in the emerging trend is

Table 1.1   Technological characteristics of Metaverse

	Emerging	*Advanced*	*Matured*
Interaction layer	Smart devices siloed apps and experiences	Immersive experiences (AR, VR, Mixed reality), Multimodal interfaces	Advanced virtual assistant, smart spaces, device independent
Content layer	e-commerce, Social networks, games, sensor technologies, Internet of Things development tools	Tokens and digital currencies, environmental mapping, Digital experience platforms (DXPs), Geo-posed and persistent data	Spatial data integration, Graph technologies
Infrastructure layer	Wireless connectivity, Web3, multiplayer platform	Blockchain, Edge cloud services, digital spatial protocols	High-bandwidth/ low-latency networking, spatial registries, P2P services, interoperability frameworks

Source Gartner (2022c)

expected. The concept of Metaverse is made possible by the convergence of different technologies. This includes more advanced and affordable AR and VR headsets that greatly enhance the user experience.

Additionally, the emergence of blockchain technology has enabled the use of NFTs and digital currencies. The ongoing COVID-19 pandemic has also contributed to the normalization of persistent and multi-purpose online communication and engagement, accelerating the trend towards digitalization. Overall, a combination of technological, economic, and social factors has led to the widespread interest and excitement around the Metaverse.

When one thinks about the economics of Metaverse, also known as Metanomics, there is a presence of opportunities in approximately all market areas. Imagine someone has an online avatar and wishes to change their clothing. In that case, they can purchase unique digitally branded garments from a virtual showroom that they have selected while browsing (Wang et al., 2022a, 2022b, 2022c, 2022d, 2022e). Alternatively, someone may want to start a small online business, like an art gallery to showcase their latest collections or create a virtual private club for visitors. The development of the Metaverse is expected to bring about new economic opportunities and a broader social impact. The exact form that the Metaverse will take is expected to evolve gradually, becoming clearer only once it reaches a significant level of adoption. In India, Tardi-Verse, a start-up based in Chennai, has made headlines for facilitating the country's first Metaverse wedding. The company, which is run by Vignesh Selvaraj and utilizes blockchain technology from Polygon, plans to host a virtual wedding reception for a couple in the Metaverse. The event will be accessible to the couple's friends and family from anywhere in the world and they will even be able to select their own digital avatars (Brabus, 2022). As per XROM, a Mumbai-based extended reality venture, there are more than 2000 AR and VR start-ups prevailing in India that will drive towards high economic growth.

1.5   Why Metaverse Continuum?

The Metaverse Continuum refers to the spectrum of digitally enhanced business models and reality worlds, which further defines how the world operates, works, and interacts. It is seen that businesses are racing towards the future in a completely different manner in what they were designed for operations, while technologies, such as extended reality, blockchain, and edge computing are converging for reshaping human experiences. Metaverse can be seen as an expanding and evolving continuum of technologies including AR, VR, design tools, digital assets, and apps driving new experiences, underpinned by connectivity technology such as cloud and 5G. Collectively, these tools and concepts are increasingly blurring the boundaries between physical and digital and show that they have the potential to revolutionize processes, boost overall operations maturity, and drive-up efficiency like never before. Accenture has mentioned that the Metaverse can be noticed as an expanding continuum in multiple directions including (Accenture, 2022a):

- Multiple technologies such as extended reality, digital twins, blockchain, AI, and smart objects including factories and cars and edge computing (Accenture, 2022a).
- Encompassing the "virt-real" which consists of a range of experiences beginning from purely virtual to the amalgamation of both physical and virtual.

- Describing the emerging spectrum of emerging consumer experiences and the business applications and models across the enterprises that will be transformed and reimagined.

As per the research conducted by Gartner, 98% of business executives believe that continuous advancements in technology are more reliable as compared to political, economic, and social trends for forming the long-term strategy of organizations. The way the internet has evolved beyond simple websites, which underpins the majority of the businesses today, it would be unfair to think that the experience of the Metaverse will be limited to the digital space. Constant development in the field of AR and VR challenges the ground assumptions about businesses and technology. The world is entering a completely new business landscape that does not have any rules or expectations yet creating a clear opportunity for shaping the world for tomorrow.

Enterprises that are deploying human-like AI are not just shaping automation benefits but rather pioneering new and collaborative forms between machines and humans. Edge capabilities and smart materials are built based on the expectations of people from the physical environment (Huynh-The et al., 2022). Organizations that are selling goods in the Metaverse environment are trying to deliver different products and the new modes of commerce that they have been piloting are helping in the creation of the internet's best practices for the future. All the companies form a new world by bringing precedents and ideas to them, shaping how people will soon live. These businesses will then find opportunities to become responsible businesses in the environment.

As of now, the future that enterprises have been rocketing for has more questions than answers. Questions have been constantly revolving around the internet and technology leaders regarding how companies will sell products and conduct business, how consumers will be able to buy the offerings in the new world, how human interaction will unfold, what would the world of work look like if the organization becomes more distributed or autonomous, and how supply chain will be managed in the meta world where some cities are smart and some are not.

In various forms, the new world companies that are beginning to build have no legacy or history—showing no right way to do anything. This shows the presence of immense opportunity. However, it is also vital to note that companies constantly push the boundaries to become a part of this new world and will operate far ahead of regulations and policies.

Without realizing, some companies have already begun the adoption of a forward-looking mindset while creating the building blocks that will further become Metaverse Continuum in the future.

The Covid-19 crisis helped develop a new generation of technology leaders that fastened the process by 4 times. In the past year, leapfroggers, a special class of companies have been introduced that worked rapidly towards the implementation of digital strategies for navigating the pandemic.

According to the Accenture report, 14% of business executives believe that Covid-19 is continuing to disrupt operations and business plans of the companies. Another 86% of the executives stated that organizations have smoothly adapted to the pandemic disruption and have found the new normal (Accenture, 2022a). Companies that have adopted the disruption would join as leaders who will shape and evolve in the emerging Metaverse Continuum. However, it is only easy to speculate about its growth in the near future. Each company has its self-idea of the optimum future, but it is a gross mistake to believe that these efforts are mutually exclusive. In some cases, a company's ambitions might be conflicting but in others, they may amplify one another.

Through this vision, it explores how today's technological innovations are becoming the collective building block of the future. The future trend investigates the entire continuum, from physical to virtual to humans and machines. It identifies where ambitious enterprises can find appropriate opportunities by uprooting from today from obsolete technology and planting themselves strategically in the future.

1.6 ECONOMIC IMPLICATIONS OF METAVERSE TECHNOLOGY

A recent study suggests that the Metaverse has the potential to increase the global GDP by $3 Trillion in the next decade if its popularity grows in a manner similar to that of mobile technology (NFTically, 2022). Economists from Analysis Group, an international consulting firm, conducted a study and discovered that the expansion of the virtual world has the potential to increase Europe's economy by 1.7% or $440 billion by the next decade. The report also suggests that widespread use of the technology in 2022 could result in a 2.8% increase in the world's GDP by 2031.

In contrast to previous popular beliefs, there is no existence of virtual economy in the Metaverse. Because of this, these innovations have larger and futuristic economic impact. It is expected that the development of Metaverse can edit various present economic factors, such as specialized industries, employment, and infrastructure.

According to the model presented by the researchers, the Metaverse is expected to significantly contribute to the global GDP. The Metaverse's share of regional GDP is estimated to be 2.3% for APAC, 0.9% for Canada, 1.7% for Europe, 4.6% for India, 5% for LATAM, 6.2% for MENAT, 1.8% for SSA, and 2.3% for the USA (Arora, 2022) (Table 1.2).

While it is still at the infancy stage, the future of the Metaverse is envisioned as massive to the global economy. As per the investigation, the Asia–Pacific region would gain maximum benefit from the Metaverse if it is adopted in the year 2022 at 2.3% which equates to $1.04 trillion (€993.9 billion). The lowest contribution will be gained by Canada with only 0.9% growth which is approximately $ 20 billion in the next 10 years.

The appeal of Metaverse technology lies in its focus on a customer-centric economy. This technology can significantly enhance employees' soft skills through VR training, four times more effectively than traditional in-person classrooms. Walmart in the US has already implemented over 17,000 VR headsets for employee training, resulting in a stunning improvement. The time required for training has been reduced from 8 hours to just 15 minutes, leading to higher efficiency and shorter training

Table 1.2 Contribution to GDP from Metaverse in 10 years

Region	Assumed secular GDP growth (%)	Metaverse's share of 10th year GDP (%)	Metaverse total contribution to GDP in 2031 ($ Trillions)
APAC	4.3	2.3	$1.04
Canada	1.1	0.9	$0.02
Europe	1.5	1.7	$0.44
India	5.4	4.6	$0.24
LATAM	1.1	5	$0.32
MENAT	1.9	6.2	$0.36
SSA	1.0	1.8	$0.04
USA	1.6	2.3	$0.56
Global	2	2.8	$3.01

Source Analysis Group (2022)

time for employees. This, in turn, is expected to significantly contribute to the rapid growth of the economy.

One crucial sector that can be revolutionized by the Metaverse is manufacturing and machinery system which is extensively complex to develop. Metaverse amalgamated with AR and VR technology can be used in reducing the cost of complex industrial products and machinery. It reduces the potential and cost of human error. Specifically, it can be used for communicating instruction in a step-by-step manner.

Modelling and simulation tools in the Metaverse can help in enhancing visibility and productibility for stakeholders (Mckinsey & Company, 2022). An "Enterprise Metaverse System" allows for the management of labour inputs, virtual inventory tracking, and cost simulations by replicating the physical world and providing improved control over implementation, design, and execution.

Additionally, the emergence of decentralized finance in the Metaverse is predicted to completely revolutionize traditional finance. The power will shift from financial institutions and centralized banks to end users. Lending and borrowing will become more streamlined, transparent, and direct. The Metaverse will allow for investment, payment, trading, and lending, ultimately rendering brick-and-mortar banks obsolete as customer interactions and services transition to this virtual world.

It is also expected that Metaverse will optimize urban development, with several cities, such as Saudi Arabia and Singapore which are already creating digital twins for improving urban planning and operational efficiencies. Digital twin helps in enabling users to visualize a 3D set-up of how the city will evolve and develop with the expansion of country population.

Some analysts also predict that the Metaverse economy, which has been developed on blockchain-based enabling technologies such as cryptocurrencies, NFTs, and smart contracts, is expected to generate revenue of $1 trillion by 2025 (Statista, 2022a, 2022b).

The timescales are short so it is important for the government and organizations to evaluate how the economic activities can be taxed in the virtual world. Some legislations that already cover e-commerce can also be adapted but in multiple areas, there is a requirement for the development of new laws under the public policy to streamline the process. It is crucial for organizations to navigate the complex international tax environment in which different jurisdictions apply varying tax treatments to digital assets.

Current discussions surrounding the Metaverse suggest that many elements need to be brought together, particularly the reorganization of local economies that may be a result of the Metaverse's growth (IBM, 2022). Metaverse will act as a haven to inspire trust and loyalty in industries. It is expected that the Metaverse will ensure boosting global economic development, and increase in purchases and languages as the lines between virtual and real get blurred. A virtual economy will help in preserving the experiences of society with digital professions that will bring profound value to the economy.

Despite the current criticism and excitement, the Metaverse has the ability to transform the current world, although in a digital manner. It is clear that it will offer more opportunities for revenue growth in the future and will benefit the expanding global economy.

References

Accenture. (2022a). *Government enters the metaverse.* https://www.accenture. com/content/dam/accenture/final/industry/public-service/document/Acc enture-Federal-Technology-Vision-2022-Government-Enters-the-Metaverse New.pdf#zoom=40. Accessed 19 Dec 2022.

Accenture. (2022b). *Protecting and serving in the metaverse continuum.* https:// www.accenture.com/us-en/blogs/voices-public-service/public-safety-tech-vision. Accessed 20 Dec 2022.

Ahn, S. J., Kim, J., & Kim, J. (2022). The bifold triadic relationships framework: A theoretical primer for advertising research in the metaverse. *Journal of Advertising, 51*(5), 592–607.

Analysis Group. (2022). *The potential global economic impact of the metaverse.* https://www.analysisgroup.com/globalassets/insights/publishing/2022-the-potential-global-economic-impact-of-the-metaverse.pdf. Accessed 8 Nov 2022.

Arora, S. (2022). How the metaverse accelerates economic development for emerging economies. *Economic Times.* https://economictimes.indiatimes. com/markets/cryptocurrency/how-the-metaverse-accelerates-economic-dev elopment-for-emerging-economies/articleshow/93375549.cms?utm_source= contentofinterest&utm_medium=text&utm_campaign=cppst. Accessed 8 Oct 2022.

Bae, J. (2021). Introduction of project-based advanced convergence structure education using metaverse in the era of future education. *Journal of Korean Association for Spatial Structures, 21*(4), 4–9.

Brabus, P (2022). *India's first metaverse marriage scheduled on February 6th in Tardi World*. News. https://thevrsoldier.com/first-metaverse-marriage-schedu led-in-tardiworld/. Accessed 17 Oct 2022.

Darbinyan. (2022). *Virtual shopping in the metaverse: What is it and how will AI make it work*. Innovation. Forbes. https://www.forbes.com/sites/forbes techcouncil/2022/03/16/virtual-shopping-in-the-metaverse-what-is-it-and-how-will-ai-make-it-work/?sh=47e1a7065f27. Accessed 15 Oct 2022.

EY. (2022). *Insights on the metaverse and the future of gaming*. https://www.ey.com/en_us/tmt/what-s-possible-for-the-gaming-industry-in-the-next-dimens ion/chapter-3-insights-on-the-metaverse-and-the-future-of-gaming. Accessed 17 Oct 2022.

Forbes. (2022a). The challenges and opportunities with the metaverse. https://www.forbes.com/sites/forbestechcouncil/2022/05/17/the-challenges-and-opportunities-with-the-metaverse/?sh=38834232495f. Accessed 3 Jan 2022.

Forbes. (2022b). *The future of social media in the metaverse*. Enterprise Tech. https://www.forbes.com/sites/bernardmarr/2022/08/24/the-future-of-social-media-in-the-metaverse/?sh=3e4e24011023. Accessed 6 Oct 2022.

Gadekallu, T. R., Huynh-The, T., Wang, W., Yenduri, G., Ranaweera, P., Pham, Q. V., …, Liyanage, M. (2022). Blockchain for the metaverse: A review. *arXiv preprint*, arXiv:2203.09738.

Gartner. (2022a). *Gartner predicts 25% of people will spend at least one hour per day in the metaverse by 2026*. Press Release. https://www.gartner.com/en/ newsroom/press-releases/2022-02-07-gartner-predicts-25-percent-of-people-will-spend-at-least-one-hour-per-day-in-the-metaverse-by-2026. Accessed 6 Oct 2022.

Gartner. (2022b). *What is a metaverse? And should you be buying in?* Information Technology. https://www.gartner.com/en/articles/what-is-a-metaverse. Accessed 17 Oct 2022.

Gartner. (2022c). *Metaverse evolution will be phased; here's what it means for tech product strategy*. https://www.gartner.com/en/articles/metaverse-evolut ion-will-be-phased-here-s-what-it-means-for-tech-product-strategy. Accessed 6 Oct 2022.

Gartner. (2022d). *Gartner predicts 25% of people will spend at least one hour per day in the metaverse by 2026*. https://www.gartner.com/en/newsroom/ press-releases/2022-02-07-gartner-predicts-25-percent-of-people-will-spend-at-least-one-hour-per-day-in-the-metaverse-by-2026. Accessed 4 Jan 2022.

Gartner. (2022e). *Gartner predicts 90% of current enterprise blockchain platform implementations will require replacement by 2021*. Newsroom. https://www.gartner.com/en/newsroom/press-releases/2019-07-03-gartner-predicts-90--of-current-enterprise-blockchain. Accessed 8 Oct 2022.

Gartner. (2022f). *What is new in the 2022 Gartner hype cycle for emerging technologies.* https://www.gartner.co.uk/en/articles/what-s-new-in-the-2022-gartner-hype-cycle-for-emerging-technologies. Accessed 3 Jan 2022.

Goldberg, Y. (2016). A primer on neural network models for natural language processing. *Journal of Artificial Intelligence Research, 57,* 345–420.

Hawkins, M. (2022a). Metaverse live shopping analytics: Retail data measurement tools, computer vision and deep learning algorithms, and decision intelligence and modeling. *Journal of Self-Governance & Management Economics, 10*(2).

Huynh-The, T., Pham, Q. V., Pham, X. Q., Nguyen, T. T., Han, Z., & Kim, D. S. (2022). Artificial intelligence for the metaverse: A survey. *arXiv preprint,* arXiv:2202.10336.

IBM. (2022). *Blockchain success starts here.* https://www.ibm.com/in-en/topics/what-is-blockchain. Accessed 12 Oct 2022.

Joy, A., Zhu, Y., Peña, C., & Brouard, M. (2022). Digital future of luxury brands: Metaverse, digital fashion, and non-fungible tokens. *Strategic Change, 31*(3), 337–343.

Lim. (2022). *70% of virtual store visitors made a purchase, new study reveals.* The Industry Fashion. https://www.theindustry.fashion/70-of-virtual-store-visitors-made-a-purchase-new-study-reveals/. Accessed 15 Oct 2022.

López-Belmonte, J., Pozo-Sánchez, S., Lampropoulos, G., & Moreno-Guerrero, A. J. (2022). Design and validation of a questionnaire for the evaluation of educational experiences in the metaverse in Spanish students (METAEDU). *Heliyon, 8*(11), e11364.

Majerová, J., & Pera, A. (2022). Haptic and biometric sensor technologies, spatio-temporal fusion algorithms, and virtual navigation tools in the decentralized and interconnected metaverse. *Review of Contemporary Philosophy, 21,* 105–121.

Mckinsey & Company. (2022). *Value creation in the metaverse.* https://www.mckinsey.com/~/media/mckinsey/business%20functions/marketing%20and%20sales/our%20insights/value%20creation%20in%20the%20metaverse/Value-creation-in-the-metaverse.pdf. Accessed 8 Oct 2022.

Nevelsteen, K. J. (2018). Virtual world, defined from a technological perspective and applied to video games, mixed reality, and the metaverse. *Computer Animation and Virtual Worlds, 29*(1), e1752.

NFTically. (2022). *How will metaverse impact the global economy.* https://www.nftically.com/blog/how-will-metaverse-impact-the-global-economy/. Accessed 18 Oct 2022.

Ooi, B. C., Tan, K. L., Tung, A., Chen, G., Shou, M. Z., Xiao, X., & Zhang, M. (2022). Sense the physical, walkthrough the virtual, manage the metaverse: A data-centric perspective. *arXiv preprint.* arXiv:2206.10326

ReportLinker. (2022). *Global Non-fungible Token (NFT) market 2022–2026. Advanced IT Market Trends.* https://www.reportlinker.com/p06268966/Global-Non-fungible-Token-NFT-Market.html. Accessed 17 Oct 2022.

Statista. (2022a). *Global digital twin market share in 2020, by industry. Hardware.* https://www.statista.com/statistics/1296192/global-digital-twin-market-share-by-industry/. Accessed 8 Oct 2022.

Statista. (2022b). *Worldwide spending on blockchain solutions from 2017 to 2024. Software.* https://www.statista.com/statistics/800426/worldwide-blockchain-solutions-spending/. Accessed 18 Oct 2022.

Time. (2021). *Why TIME is launching a new newsletter on the metaverse.* https://time.com/6118513/into-the-metaverse-time-newsletter/. Accessed 5 Oct 2022.

Vidal-Tomás, D. (2022). The new crypto niche: NFTs, play-to-earn, and metaverse tokens. *Finance Research Letters,* 102742.

Wang, F. Y. (2022). Metavehicles in the metaverse: Moving to a new phase for intelligent vehicles and smart mobility. *IEEE Transactions on Intelligent Vehicles, 7*(1), 1–5.

Wang, G., Badal, A., Jia, X., Maltz, J. S., Mueller, K., Myers, K. J., ..., Zeng, R. (2022). Development of metaverse for intelligent healthcare. *Nature Machine Intelligence, 4*(11), 922–929.

Wang, H., Chen, D., & Deng, Q. (2022). The formation, development and research prospect of educational metaverse. *Education Journal, 11*(5), 260–266.

Wang, M., Yu, H., Bell, Z., & Chu, X. (2022). Constructing an edu-metaverse ecosystem: A new and innovative framework. *IEEE Transactions on Learning Technologies.*

Wang, X., Wang, J., Wu, C., Xu, S., & Ma, W. (2022). Engineering brain: Metaverse for future engineering. *AI in Civil Engineering, 1*(1), 1–18.

Wang, Y., Su, Z., Zhang, N., Xing, R., Liu, D., Luan, T. H., & Shen, X. (2022). A survey on metaverse: Fundamentals, security, and privacy. *IEEE Communications Surveys & Tutorials.*

XR Today. (2022a). Artificial intelligence in the metaverse: Bridging the virtual and real. *Virtual Reality.* Available at https://www.xrtoday.com/virtual-reality/artificial-intelligence-in-the-metaverse-bridging-the-virtual-and-real/. Accessed 18 Oct 2022.

XR Today. (2022b). *Meta quest pro hits store shelves.* https://www.xrtoday.com/mixed-reality/meta-quest-pro-hits-store-shelves/. Accessed 6 Oct 2022.

Youtube. (2021). *The metaverse and how we'll build it together—Connect 2021.* https://www.youtube.com/watch?v=Uvufun6xer8. Accessed 17 Oct 2022.

Zhao, Y., Jiang, J., Chen, Y., Liu, R., Yang, Y., Xue, X., & Chen, S. (2022). Metaverse: Perspectives from graphics, interactions and visualization. *Visual Informatics, 6*(1), 56–67.

CHAPTER 2

Metaverse in the Technological World

Abstract In the technological set-up, the changes experienced by the metaverse are crucial and need emphasis to understand the concept in a well-defined manner. Blockchain technology has acted as a building block of the metaverse that exists to track orders, payments, accounts, and production. Also, digital twinning acts as a virtual model of the product and service which is also becoming an essential block of the metaverse. Apart from the above mentioned concepts, this chapter also discusses a multiverse which is a hypothetical collection of different virtual worlds. Though its existence is questionable, its discussion becomes important while discussing the virtual world. Hyperautomation is another concept that needs exposure along with the metaverse that this chapter covers in brief.

Keywords Metaverse · Web 2.0 · Web 3.0 · Digital twinning · Multiverse · Artificial intelligence · Internet of Everything · Hyperautomation

© The Author(s), under exclusive license to Springer Nature
Singapore Pte Ltd. 2023
R. Gupta and S. K. Pal, *Introduction to Metaverse*,
https://doi.org/10.1007/978-981-99-7397-2_2

2.1   METAVERSE VS WEB 2.0 VS WEB 3.0

The initial stage of the internet, often referred to as the first generation, encompasses the time frame from the late 90s to the early 2000s, during which websites were primarily composed of static content and individuals were primarily consuming information. The advent of Web 2.0 brought about the ability for users to generate their own content and gave rise to social networking sites. Before deep diving into the details of the metaverse, it is crucial to lay the foundation of the features of Web 2.0, i.e., today's metaverse and Web 3.0, i.e., emerging characteristics of the metaverse (Grayscale Research, 2022) (Table 2.1).

In the current Web 2.0 form, it limits developers in several ways. For instance:

- Web 2.0 lacks interoperability which generates the requirement of separate software written based on the standards of each platform.
- Web 2.0 lacks composability and most of the programmes on this web do not "plug into" each other just like Lego pieces without APIs (JP Morgan, 2022).
- Secure a permanent revenue share with the platform where the developer's game is hosted (e.g., Valve).

From the standpoint of the user, the following limitations can be noticed:

- Web 2.0 lacks portability where users are not able to port their avatars, personas, data, achievements, and list of friends elsewhere.
- Web 2.0 lacks monetization opportunities where users are not able to monetize the achievements of the characters and have no character creation capabilities.
- Users cannot participate without getting authorization from the intermediary or central party.

While the Web 2.0 model might gather influence in the short run, developers are sure that it will lose its significance with time, unless it embraces a newly developing "user-controlled" model. Web 3.0 is an emerging architecture of the metaverse, virtual worlds are ecosystems where products and ideas are openly sourced, released, and monetized, independently by the community of creators, developers, players, users,

Table 2.1   Web 2.0 versus Web 3.0 approach to Metaverse

		Web 2.0	*Web 3.0*
Platform characteristics	Example virtual worlds	• Second life • Roblox • Fortnite • World of Warcraft	• Decentraland • The Sandbox • Somnium space • Cryptovoxels
	Organizational structure	• Centrally owned • Decisions are based on adding shareholder value	• Community governed generally through a foundation of decentralized autonomous organizations • Native tokens are issued and enable participation in governance • Decisions are based on user consensus
	Data storage	Centralized	Decentralized (game assets)
	Platform format	• PC/Console • VR and AR • Mobile/Applications	• PC/Console • VR and AR • Mobile/ Applications (coming soon)
	Payment infrastructure	Traditional payments, such as credit and debit card	Crypto wallets
User interactions	Digital assets ownership	Leased within platform, where purchased	Owned through NFTs
	Digital assets portability	Locked within platform	Transferable
	Content creators	Game studios and developers	• Community • Game studios and developers
	Activities	• Socialization • Multiplayer games • Game streaming • Competitive games (e.g., E-ports)	• Play to earn games • Experiences • (Others are same as Web 2.0)

(continued)

Table 2.1   (continued)

		Web 2.0	Web 3.0
	Identity	In-platform avatar	• Self-Sovereign and interoperable identity • Anonymous private-key-based identities
Commercials	Payments	In-platform virtual currency (e.g., Roblox and Robux)	Cryptocurrencies and tokens
	Content revenues	Platforms or App stores earn 30% of every game purchased; 70% goes to the developer	• Peer to peer; developers (content creators) directly earn revenue from sales • Users/Gamers can earn through participation or platform governance • Royalties or secondary trades of NFTs to the creator

Source JP Morgan (2022)

and other ecosystem members. Therefore, these virtual worlds operate as autonomous, self-sufficient economic systems where the inhabitants of the virtual world, instead of the platforms, regulate the generation and distribution of value.

The potential of Web 3.0 is immense opportunities which can go beyond streaming, social media, and online shopping. Capabilities, such as AI, machine learning, and the Semantic web, considered some of the core of 3.0, have the potential to increase application in new areas and bring vast improvement in user interaction (Christensen & Robinson, 2022). Some core features of Web 3.0, such as permissionless and decentralization systems have the capability to give more control to the users over personal data. This could also aid in curtailing data harvesting, which involves collecting information from web users without their permission or compensation. This will further reduce the network effects that have allowed technology giants to gain near-monopolistic power through manipulative advertising and marketing tactics.

In contrast, decentralization comes with significant regulatory and legal risks. Cybercrime, hate speech, and misinformation are already difficult to assess and it is expected that the process will become even more difficult in central control. Also, the decentralized web would make enforcement and regulation very difficult. For example, which national laws will govern a website whose content is hosted in multiple countries around the world? If Web 1.0 could be compared to the era of black-and-white movies, then Web 2.0 would represent the age of colour/ simple 3D, and Web 3.0 would entail immersive experiences in the virtual universe.

2.2   BLOCKCHAIN TECHNOLOGY

Blockchain is an immutable and shared ledger that facilitates tracking assets and recording transactions in the business network. In a virtual sense, anything of value can be exchanged and monitored on a blockchain network, thereby reducing risk and minimizing expenses for those involved.

Blockchain becomes important because business functions on information. The faster the process becomes, the more accurate the results are. The blockchain is well-suited for transmitting information due to its ability to store data on an unchangeable ledger, accessible only by authorized network members. Tracking orders, accounts, payments, and production is made easier with the use of a blockchain network. With a shared view of the truth, all transaction details can be seen from start to finish, providing greater assurance and opportunities for increased effectiveness. This transparency can lead to improved confidence and efficiency (IBM, 2022) (Fig. 2.1).

The use of blockchain technology is predicted to enable metaverse businesses to provide customers with integrated services that combine their physical presence with digital 3D presences. This will transform how customers interact with and trade cryptocurrencies or unique digital assets such as NFTs. By converting virtual goods, digital art, and personal experiences into secure NFTs and storing them as assets on the metaverse blockchain, users can establish a self-sustaining digital economy. These

Blockchain Plans

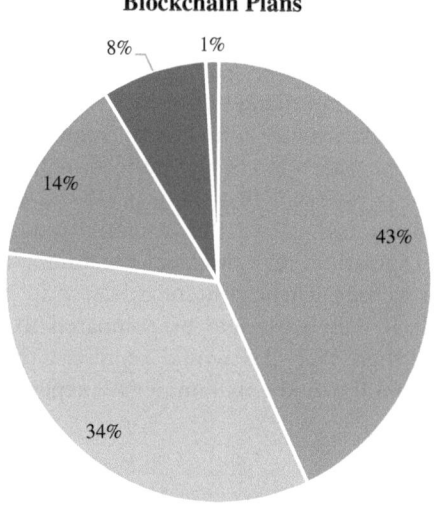

- On the Radar but no actions planned
- In medium or long term planning
- Have already invested and deployed
- No Interest
- In short term planning / actively experimenting

Fig. 2.1 Blockchain plans by enterprises (*Source* IBM, 2022)

NFTs can be exchanged for cryptocurrencies to purchase other metaverse entities or can be cashed out for fiat money at any time. According to Statista, the expenditure on blockchain-related solutions is expected to reach 19 billion USD by 2024 (Statista, 2022a) (Table 2.2).

Blockchain has persuasive and clear roles in developing the future when only one firm does not control the metaverse and multiple platforms supply a variety of information and experiences. The ability for users to move within the metaverse with their avatars and valuable digital assets is made possible by the platform mobility offered by blockchain technology. Cryptocurrency could provide a local means of payment within the metaverse, facilitating fast, direct transactions between individuals without intermediaries. The European Union's GDPR mandates a certain level of portability for metaverse platforms. Blockchain can serve as a bridge between platforms, keeping track of the ownership of specific digital assets

Table 2.2 Blockchain-based virtual economy segments

Self-custody and access—wallets/front end applications

Agents

Decentralized Finance (DeFi)	*NFTs-sovereign virtual goods*	*Decentralized governance*	*Decentralized cloud services*	*Self-sovereign identity*
• Aggregators • DeFi primitives • Oracles • Data • Marketplaces • Unit of value-"Internet Money"	• Minting houses • Marketplace token standards • Metadata standards • Hybrid NFT+FT • Physical redeemable NFTs	• DAO frameworks • Voting mechanisms • Staking and slashing • Multigeniture wallets • Community audits	• Storage • Compute • Databases • Query and APIs	• Decentralized Identifiers (DIDs) • Verifiable claims • Creator coins

Programmability layer
Transaction layer
Peer-to-peer networks

Source Gray Scale (2021)

linked to identities. Its effectiveness has already been demonstrated in the NFT marketplace (Premium, 2022). For instance, Axie Infinity users can opt for purchasing an NFT on the Axie Marketplace, increase its worth by training it in the game, and sell it on OpenSea.

Due to its recent emergence, opinions on the potential of blockchain technology are mixed. A TechRepublic Research survey found that 70% of professionals claimed to have never used blockchain technology. Nonetheless, 64% anticipate that blockchain will affect their industry, and most of them are optimistic about the outcomes.

Gartner's Report reflects that the corporate value provided by blockchain will be slightly over $360 billion by 2026, increasing to more than $3.1 trillion by 2030 (Gartner, 2022a). Blockchain technology can focus on preventing tampering, keeping personal data safe, and allowing parties to validate the legitimacy of a file.

2.3   DIGITAL TWIN

There is a lot to talk about metaverse, thanks to the continuous research and focus on the developers towards its growth. The next advancement in technology involves the implementation of digital twin in the Internet of Things and the metaverse, which serves to further diminish the divide between humans and the digital world. As the metaverse gains popularity, with the adoption of several companies, it can be observed that it is primarily utilized in three main areas.

- Creating Avatars
- Help scanning objects with photogrammetry
- Establishing the association between digital twins with live data

This section focuses on the third application of the metaverse as we explore how digital twins relate to one another and fit together.

Digital twin refers to a virtual model of a product, process, and service. Alternatively, it considers a real-life object and develops an exact copy of the same object in the digital world. It is an essential building block of the metaverse. Consider the possibility of visiting the virtual store of a fashion e-commerce company, where you can try on clothes prior to buying them. An ideal solution is to use a digital twin avatar to test the clothing and ensure that it fits a user's actual measurements.

Conducting a productive professional meeting in a metaverse-based meeting room requires interaction with an accurate replica of the company's instruments, equipment, and information system by virtual meeting participants. Similarly, a technical training programme conducted in a metaverse will be enhanced if technicians can manipulate 3D models of complex systems. Digital twin technology can turn these concepts into reality, facilitating the creation of a more realistic metaverse. With the aid of simulation technology and digital twin, the metaverse can offer remote maintenance workshops for machinery that requires servicing and may also be integrated with or mapped onto an actual workshop. These integral characteristics demonstrate the fundamental role of digital twins in the metaverse.

It is important to evaluate how digital twin works in the metaverse and how they can be incorporated into the metaverse.

Product: Digital twins are used in designing the product.

Production: Digital twins are used in the production process to validate the production process.

Performance: The digital twin for performance purposes acquires data from products in operation, and assesses the information, offering practical guidance to enable informed decision-making.

The combination and integration of the three digital twin types are collectively known as digital threads, which can be incorporated into various products by gathering data at each product and production stage. Developers and engineers use manufacturing, physical, and operational data to form a digital twin. This information, along with an AI algorithm, is integrated into a virtual model based on physics. Analytics applied to these models can provide valuable insights regarding the physical asset. The continuous flow of data allows for a thorough analysis, turning the digital twin into a live model of the physical equipment. The digital twin market was valued at USD 7.48 billion in 2021 and is expected to grow at a CAGR of 39.1% from 2022 to 2030 (Fig. 2.2).

As per Statista, more than 22% market share was attributed to the manufacturing industry in 2020 followed by the automotive industry with a market share of 18% in the same year.

Digital twins have the potential to improve the metaverse infrastructure at a global scale by enabling the analysis of digital and virtual counterparts

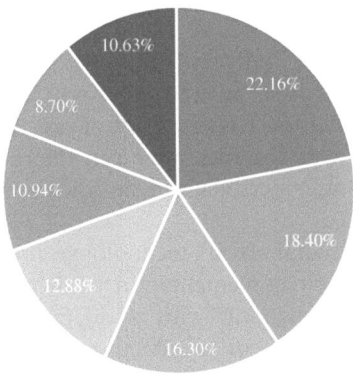

* Manufacturing * Automotive * Aviation * Energy and Utilities * Healthcare * Logistics and Retail * Others

Fig. 2.2 Industry wise distribution of Global Digital Twin Market in 2020 (*Source* Statista, 2022)

to determine and predict the current and future state of physical assets. Integrating digital twins in the metaverse can provide enterprises and organizations with better insights into product performance and enhance customer service. Various sectors stand to benefit from the application of digital twins in the metaverse. Some of those sectors are as follows:

Manufacturing: Virtual copies of the plants and factories initiate a transparent production process. The digital twin is expected to revolutionize the manufacturing industry in the Metaverse by affecting the way products are designed, produced, and serviced, resulting in improved efficiency and optimization and reduced lead times.

Automobile: Digital twins in the automobile sector can help in virtually creating a virtual model of a physically connected vehicle. It captures functional and behavioural data of the vehicle and further assists in analysing the overall performance and connected features of the vehicles. Digital twins help in delivering personalized customer service for the customers. It is expected that in future, the metaverse can act as a platform for virtual showrooms, and automobile expos and their digital twins can help users have real experiences with automobiles.

Retail: Digital twins applied in the metaverse can play a crucial role in mounting the customer experience by creating 3D virtual models of products and showrooms, delivering customers a real-life experience in the comfort of their homes. Digital twin helps in security implementation, better store planning, energy management, and security implementation in a well-optimized manner.

Healthcare: The medical industry has already experienced advantages from using digital twins in various areas such as surgery training, organ donation, and decreasing risks in medical procedures. Integrating digital twin technology with the metaverse can improve the process of patient monitoring and offer personalized healthcare, including preventive measures.

Smart Cities: 3D digital twin technology already exists. One convincing example of such implementation is virtual Singapore. It is expected that smart city planning with a digital twin in the metaverse can ensure the enhancement of effective management of human resources, reduced ecological footprints, and economic development (Leeway Hertz, 2022a). The idea would be to improve the quality of life of the citizens in both the virtual and physical worlds.

Industries IoT: By integrating digital twins into the metaverse, industrial companies can digitally track, monitor, and control their industrial systems. These digital replicas capture operational data, including environmental factors such as configuration, financial models, and location, to predict future industrial operations.

2.4 Metaverse vs Multiverse

In literal terms, multiverse means multiple universes. In definitive words, multiverse refers to the collection of digital space with unique traits and characteristics. These digital spaces are completely separate and different. Therefore, people can also learn, interact, play, and do more in the multiverse. The difference is that these activities are done on isolated ecosystems, instead of only one platform. Multiplayer video games, social media platforms, and online games are examples of digital space that are already a part of the multiverse.

Unlike the metaverse, the multiverse refers to various universes instead of just one. Scientists believe the multiverse is real, meaning it exists beyond our known universe's boundaries. This theory is based on evidence from experiments that have shown that quantum physics can create multiple copies of an object or event. This concept only works in theory.

Although the concept of the multiverse seems to be quite achievable and appealing, it is still in a hypothetical state. For the basis of the multiverse, the scientific explanations are evident in the string theory which tries to explain the smallest unit of matter. String theory (Douglas, 2019) says that there is a presence of several universes with separate collections of rules and each universe is different from the other. Unfortunately, the theory could not explain the universe without assuming that the other universe also exists today.

Imagine yourself seeing the internet on-screen and the minute you put on your VR headsets, the internet goes all around you. This is known as Metaverse. On the other hand, the multiverse is the theory of a parallel universe as it exists in the world of physics, astronomy, cosmology, and other related fields of study. The metaverse represents the other ecosystems of the virtual world that are disassociated. However, these ecosystems are present in the multiverse, metaverse does not allow users to transition seamlessly between different worlds (Table 2.3).

Comparing the metaverse and multiverse, it is evident that both of them have significant differences. The metaverse serves as the foundation for a connected virtual world in the future, while the multiverse presents

Table 2.3 Difference between Metaverse and Multiverse

	Metaverse	*Multiverse*
Concept	Networks of 3D virtual worlds where users can play, work, hop, and connect	Hypothetical virtual world collection
Technology	AR, VR, Artificial Intelligence (AI), and other virtual platforms	2D components present
Interconnectivity	Metaverse is universal and whole and endures interoperability	Due to the presence of multiplicity, there is the presence of multiple worlds. These worlds might or might not be interoperable
Status	Presently, at the buildable stage and developers and researchers are still exploring its implementation	It is questionable if it exists or not as the concept does not follow any concrete creation process
Architecture	Shared and connected world with a specific order for informational flow	Random, containing multiple virtual worlds without any particular order or flow
Entitles	The participants are inclusive of robots, AI humans, and digital avatars (Abrol, 2022)	Virtual worlds are the only entities that make up the Multiverse
Platforms	Creation in platforms by tech giants, such as Microsoft, Meta, and others	Horizon, Roblox, Fortnite, and many more platforms
Ecosystems	One; entirely connected virtual world	Completely different and disconnected ecosystem of virtual worlds

the possibility of multiple digital ecosystems. The metaverse takes the lead over the multiverse in terms of a unified user interface, but the metaverse is still in the developmental stage, leaving room for further expansion.

The metaverse is generally considered superior to the multiverse due to its cohesive user interface. However, it remains to be seen what will happen as the metaverse continues to grow and develop. Despite common assumptions, the multiverse already contains numerous practical virtual environments for real-life applications.

The multiverse currently contains virtual worlds that serve various purposes such as work, gaming, and social interaction, making it difficult to determine which is better between the metaverse and the multiverse.

2.5   ARTIFICIAL INTELLIGENCE (AI)

As per the research of Weforum, the AI market is constantly expanding with a CAGR of 38.1% between 2022 and 2030. It is expected that as many as 97 million people will work in the AI space by 2025 (Russo, 2020). In 2018, the global AI software market revenue was $10.1 billion which is expected to grow exponentially to $126 billion by 2025. The AI market encompasses various applications, such as machine learning (ML), natural language processing (NLP), computer vision (CV), and robotic process automation (RPA) (Fig. 2.3).

Big tech companies have made significant investments in AI through acquisitions and research and development. Companies like Google, Microsoft, IBM, and Samsung have submitted thousands of AI patent applications, and start-ups related to AI are receiving billions of dollars in funding annually.

In 2022, AI adoption is progressing steadily, with 35% of companies reporting using AI in their operations, a four-point increase from 2021. However, the rate of AI adoption varies depending on the geography, company, and industry. Companies in China and India are at the forefront, with nearly 60% of IT professionals in those countries reporting their organizations actively utilizing AI, a significantly higher adoption rate compared to countries such as South Korea (22%), Australia (24%), the USA (25%), and the UK (26%) (Fig. 2.4).

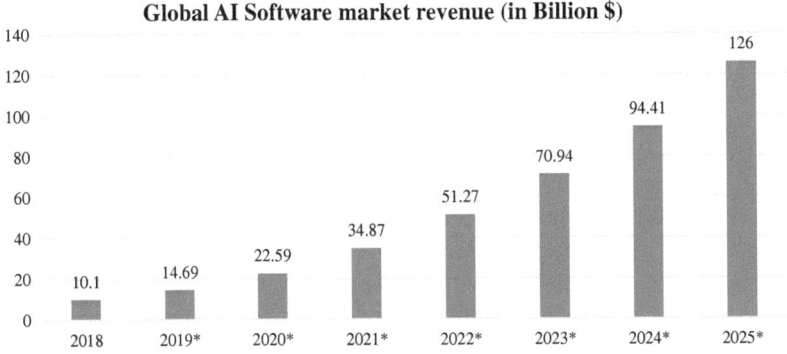

Fig. 2.3   Global AI software market revenue (*Source* World Economic Forum, 2018)

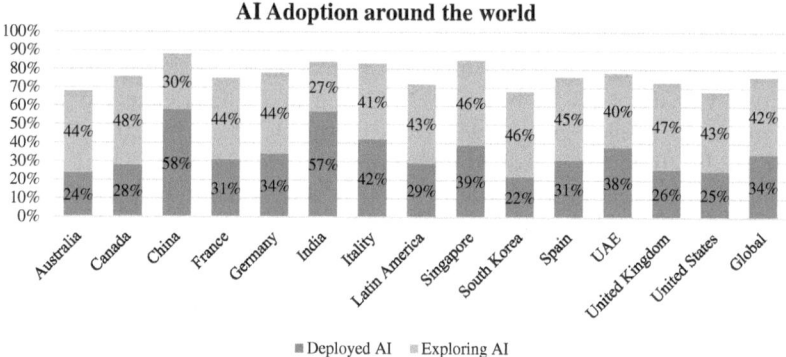

Fig. 2.4   AI adoption around the world (*Source* Watson, 2022)

Perhaps, AI is one of the strong concepts that will empower technologies like metaverse and automation at scale as compared to other concepts. The application of deep learning-based software will lead to the independent operation of chatbots and other NLP platforms. These AI systems will be able to interpret and respond to images, videos, and text in various languages without human intervention. Generating high-quality animation, 3D images, visuals, speech, and metaverse artwork requires extensive training data and modelling. AI will play a crucial role in the creation of these elements (Watson, 2022). AI will be utilized to automate virtual transactions through smart contracts, blockchain technologies, and decentralized ledgers.

Textually, the use of AI is immense. However, it is still important to assess the use-case where AI can portray itself as an indispensable part of the metaverse.

Creating accurate Avatars: The user experience is central to the metaverse, and the quality of the avatar will have a significant impact on that experience. AI has the ability to create realistic and expressive avatars by analysing 2D images or 3D scans of users, adding features such as facial expressions, hairstyles, emotions, and age-related changes to make the avatar more dynamic. Companies like "Ready Player Me" are already utilizing AI to create avatars for the metaverse, and "Meta" is developing its own version of this technology (XR Today, 2022a).

Digital humans: Digital humans are AI-enabled characters that exist in the metaverse, similar to Non-Playing Characters (NPC) in video games that can respond to user actions in the VR world. They are not replicas of actual people but are created entirely using AI technology. Digital humans have various applications in the metaverse, from NPCs in gameplay to automated assistants in VR workplaces. Companies like Soul Machines and Unreal Engine have already invested in this direction to develop this essential aspect of the metaverse.

Multilingual Accessibility: Digital humans utilize AI primarily for language processing, by converting natural languages like English into a machine-readable format, analysing them, generating a response, converting the response back into English, and sending it to the user in just a fraction of a second, similar to real conversation. AI can be trained to convert the response into any language, making the metaverse accessible to users worldwide.

VR world expansion at scale: As AI is fed with historical data, it learns and tries to generate its own output, which improves with each new input, human feedback, and machine learning reinforcement. Over time, AI will be capable of completing tasks and producing results nearly as well as humans. Nvidia is one of the companies working on this technology and training AI to create complete virtual worlds. The progress will be crucial in enhancing the scalability of the metaverse since it will become simpler to add new worlds without human involvement.

Intuitive Interfacing: AI has the ability to aid in Human–Computer Interactions (HCI) by utilizing advanced VR headsets with sensors to detect muscular and electrical patterns, allowing for accurate movement prediction within the metaverse. Additionally, AI can assist in replicating a genuine sense of touch in VR and enable voice navigation to interact with virtual objects without the use of hand controllers.

Ultimately, it is not possible to create an authentic and scalable metaverse experience without AI. That's why companies like Meta are working closely with ethics groups and think tanks to stem the risks of AI without curbing the potential of technology.

2.6 Internet of Everything (IoE)

Gartner names IoE as a transformative innovation of 2012, but it has already moved down the hype curve due to other emerging technological trends. Metaverse has suddenly become the centre of attention. And experts are calling it the "future of the internet" (Fig. 2.5).

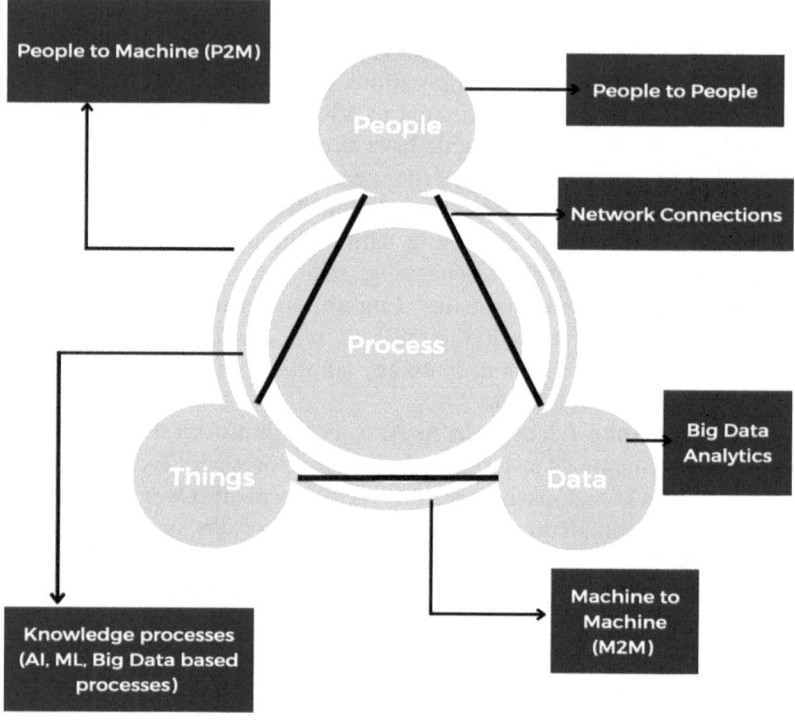

Fig. 2.5 Elements of IoE (*Source* Bandara, 2016)

IoE is a network connecting data, individuals, processes, and devices to create a more comprehensive and engaging experience. While it's only been around for two decades, it's hard to imagine life without the World Wide Web. In the past, we used to complete tasks physically without a second thought. For example, we would drive to a travel agent to book plane tickets and plan vacations, or go to a restaurant to order food and talk to friends in person. Over time, the physical world has merged with the digital world, allowing us to complete these tasks in seconds through websites, which have now been replaced by applications.

What comes next? It would be a metaverse we imagined everything to be. One will wear AR/VR headsets and immerse himself/herself in the place one wants to visit, without visiting it physically. He/she will explore a new city, review a new destination, get up close to the dishes one wants to order from a particular restaurant and engage with people as if they are right in front of them. It is expected that the metaverse will change the paradigm from IoE.

The metaverse is a faster version of IoT platforms, and its essential foundation will be provided by the data streams, resulting in a world that is more interconnected and immersive. An example of this is that renewable energy sources will depend heavily on real-time data that is transmitted to IoT platforms, which are connected to various systems such as solar panels, weather sensors, battery management systems, wind turbines, and grid systems. The metaverse will enable the management of such a massive amount of data when connected, allowing for scalability with IoT, allowing for optimized sustainable power use. Also, it is expected that with the interaction of the metaverse and IoT, enhanced productivity and safety can be attained. The use of Metaverse and IoT will provide valuable data on temperature, pressure, proximity, and operational state, which can help to enhance safety procedures by alerting people of hazardous conditions and giving them a complete 360 view of their surroundings. It is clear that the future of multidimensional experiences lies in the combination of Metaverse and IoT.

2.7   HYPERAUTOMATION

Hyperautomation refers to the technology that enables the organization to automate business processes with minimal human interventions resulting in a minimized human-induced error and improved productivity. It encompasses Robotic Process Automation (RPA), Machine Learning (ML), and AI that enables purpose-driven and outcome-oriented solutions for streamlining the processes without compromising customer satisfaction and loyalty. 80% of organizations will have hyperautomation on the technological roadmap by 2024 (Afshar, 2022).

Industries are increasingly seeking automation, but their current architecture is hindering progress. The greatest demand for automation comes from research and development (39%), administrative/operational (38%), customer service (33%), and marketing (26%) teams. Robotic Process Automation (RPA) leaders, such as UiPath are already experimenting with semantic automation in order to improve the common understanding between AI and humans by creating bots that rely on conversational programming instead of keywords or complex codes. Semantic automation can pave the path for life-like chatbots that will further populate the metaverse and greatly enhance customer experiences.

Effective collaboration between AI, systems, and people is the crowning glory of hyperautomation and has been occurring throughout business process automation for almost a decade. In a similar manner, metaverse will also combine many facets of modern technology, ML, and hyperautomation bringing together AI, process mining, RPA, human verification, and virtual collaboration into one holistic solution.

The AI/Human relationship achieved by hyperautomation yields immense benefits to both customers and businesses and is the building block of what the metaverse stands for. One will easily be able to improve enterprise efficiency and customer satisfaction while reducing costs and inaccuracies.

Fintech companies and banks are incorporating VR/AR technology in their products to keep up with technological advancements. Acorns, a fintech company, is providing an AR-based debit card for its users to view through their phones, while Westpac Banking Corporation offers budgeting and financial data visualization through AR on smartphones.

The main objective of these banking advancements is to ensure that the services are easy to use and provide convenience to users. These institutions can drive towards the next big change in the offering by leveraging hyperautomation for connecting the customer experiences to their operational infrastructure, whether through banking systems or Customer Relationship Management (CRM) for system engagement.

The combination of AR and VR features in the Metaverse, along with AI, ML, and blockchain-based DApps, among other relevant hyperautomation technologies, will create opportunities for enhanced banking experiences, including:

Multilingual Assistance: Metaverse banking services will be accessible to people worldwide in their preferred language, thanks to the assistance of AI. The process of language translation from human to machine-readable format, analysis, response generation, and re-conversion to human-readable language takes only a few seconds, enabling users to comprehend the information easily.

Highly intuitive interface: AI-enabled VR headsets with advanced sensors can improve human–computer interactions by detecting and predicting the user's muscular and electrical patterns, allowing for customized movement and use of the metaverse. In the banking sector,

VR technology can provide users with a realistic sense of touch and voice-activated navigation, eliminating the need for handheld controllers when interacting with virtual objects.

Digital Human 24/7 service: These AI-powered 3D versions in the metaverse can interact and respond to user actions, and the banking sector can use them for 24/7 automated customer assistance and other financial tasks. The integration with the bank's backend systems through hyperautomation technologies enables real-time customer service.

Hyperautomation is the idea that anything that can be automated in an organization should be automated. When mixed with Metaverse, will witness a never seen digital transformation. Thus different technologies will help in improving the implementation of metaverse and the combination of technologies will help progress the service standards by covering never heard use-cases.

References

Abrol, A. (2022). *Metaverse vs. Multiverse—What's the difference?* https://www.blockchain-council.org/metaverse/metaverse-vs-multiverse/. Accessed 8 Oct 2022.

Afshar. V. (2022). *80% of organizations will have hyperautomation on their technology roadmap by 2024.* Digital Transformation. https://www.zdnet.com/article/80-of-organizations-will-have-hyperautomation-on-their-technology-roadmap-by-2024/. Accessed 22 Oct 2022.

Bandara, I. (2016). *The evolving challenges of internet of everything: Enhancing student performance and employability in higher education.* ResearchGate. https://www.researchgate.net/figure/Four-pillar-network-connection-of-Internet-of-Everything-IoE-in-higher-education_fig2_299848797. Accessed 18 Oct 2022.

Christensen, L., & Robinson. A. (2022). *The potential global economic impact of the metaverse.* Analysis Group. https://www.analysisgroup.com/globalassets/insights/publishing/2022-the-potential-global-economic-impact-of-the-metaverse.pdf. Accessed 12 Oct 2022.

Douglas, M. R. (2019). The string theory landscape. *Universe, 5*(7), 176.

Gartner. (2022a). *Gartner predicts 25% of people will spend at least one hour per day in the metaverse by 2026.* Press Release. https://www.gartner.com/en/newsroom/press-releases/2022-02-07-gartner-predicts-25-percent-of-people-will-spend-at-least-one-hour-per-day-in-the-metaverse-by-2026. Accessed 6 Oct 2022.

Gartner. (2022b). *What is a metaverse? And should you be buying in?* Information Technology. https://www.gartner.com/en/articles/what-is-a-metaverse. Accessed 17 Oct 2022.

Gartner. (2022c). *Metaverse evolution will be phased; here's what it means for tech product strategy.* https://www.gartner.com/en/articles/metaverse-evolut ion-will-be-phased-here-s-what-it-means-for-tech-product-strategy. Accessed 6 Oct 2022.

Gartner. (2022d). *Gartner predicts 25% of people will spend at least one hour per day in the metaverse by 2026.* https://www.gartner.com/en/newsroom/ press-releases/2022-02-07-gartner-predicts-25-percent-of-people-will-spend-at-least-one-hour-per-day-in-the-metaverse-by-2026. Accessed 4 Jan 2022.

Gartner. (2022e). *Gartner predicts 90% of current enterprise blockchain platform implementations will require replacement by 2021.* Newsroom. https://www.gartner.com/en/newsroom/press-releases/2019-07-03-gartner-predicts-90--of-current-enterprise-blockchain. Accessed 8 Oct 2022.

Gartner. (2022f). *What is new in the 2022 Gartner hype cycle for emerging technologies.* https://www.gartner.co.uk/en/articles/what-s-new-in-the-2022-gar tner-hype-cycle-for-emerging-technologies. Accessed 3 Jan 2022.

Gray Scale. (2021). *The Metaverse.* https://grayscale.com/wp-content/uploads/ 2021/11/Grayscale_Metaverse_Report_Nov2021.pdf.

Grayscale Research. (2022). *The Metaverse.* Available at https://grayscale.com/ wp-content/uploads/2021/11/Grayscale_Metaverse_Report_Nov2021.pdf. Accessed 18 Oct 2022.

IBM. (2022). *Blockchain success starts here.* https://www.ibm.com/in-en/top ics/what-is-blockchain. Accessed 12 Oct 2022.

Morgan, J. P. (2022). *Opportunities in the metaverse.* Content. https://www. jpmorgan.com/content/dam/jpm/treasury-services/documents/opportuni ties-in-the-metaverse.pdf. Accessed 12 Oct 2022.

Leeway Hertz. (2022a). *Digital twin and metaverse.* https://www.leewayhertz. com/digital-twin-and-metaverse/. Accessed 8 Oct 2022.

Leeway Hertz. (2022b). *Metaverse: Uplifting the virtual gaming.* https://www. leewayhertz.com/gaming-in-metaverse/. Accessed 22 Nov 2022.

Premium. (2022). *Research: The current state and predictions for the future of blockchain in the enterprise.* https://www.techrepublic.com/resource-library/ research/research-the-current-state-and-predictions-for-the-future-of-blockc hain-in-the-enterprise/. Accessed 8 Oct 2022.

Russo. A. (2020). Recession and Automation Changes Our Future of Work, But There are Jobs Coming, Report Says, World Economic Forum. Accessed from https://www.weforum.org/press/2020/10/recession-and-automationchanges-our-future-of-work-but-there-are-jobs-coming-report-says-52c5162fce/ on 20th Dec 2021.

Statista. (2022a). *Global digital twin market share in 2020, by industry.* Hardware. https://www.statista.com/statistics/1296192/global-digital-twin-market-share-by-industry/. Accessed 8 Oct 2022.

Statista. (2022b). *Worldwide spending on blockchain solutions from 2017 to 2024.* Software. https://www.statista.com/statistics/800426/worldwide-blockchain-solutions-spending/. Accessed 18 Oct 2022.

Watson. (2022). *IBM global AI adoption index 2022.* https://www.ibm.com/downloads/cas/GVAGA3JP. Accessed 8 Oct 2022.

World Economic Forum. (2018). *Machines will do more tasks than humans by 2025 but Robot revolution will still create 58 million net new jobs in next five years.* News Releases. Available at https://www.weforum.org/press/2018/09/machines-will-do-more-tasks-than-humans-by-2025-but-robot-revolution-will-still-create-58-million-net-new-jobs-in-next-five-years/. Accessed 10 Oct 2022.

XR Today. (2022a). Artificial intelligence in the metaverse: Bridging the virtual and real. *Virtual Reality.* Available at https://www.xrtoday.com/virtual-reality/artificial-intelligence-in-the-metaverse-bridging-the-virtual-and-real/. Accessed 18 Oct 2022.

XR Today. (2022b). *Meta quest pro hits store shelves.* https://www.xrtoday.com/mixed-reality/meta-quest-pro-hits-store-shelves/. Accessed 6 Oct 2022.

CHAPTER 3

Seven Layers of Metaverse

Abstract Metaverse comprises seven fundamental layers dedicated to the framework of the structure, output, and ethical framework of an idealized decentralized metaverse. These seven layers are experience, discovery, creator economy, spatial computing, decentralization, human interface, and infrastructure. Each layer and its application have been discussed in detail to understand its contribution to the world of the metaverse.

Keywords Metaverse · Experience · Discovery · Creator Economy · Spatial Computing · Decentralization · Human Interface · Infrastructure

3.1 INTRODUCTION TO SEVEN LAYERS OF METAVERSE

The Metaverse is a highly disputed but fascinating subject of the present era, and when Mark Zuckerberg recently renamed Facebook to Meta, it caused a stir on social media. In October, Google recorded around 2.62 million searches related to this topic. (European Gaming, 2021).

In the past 2 years, metaverse technologies have matured and attracted huge investment from firms such as Meta platforms, Microsoft, Epic Games, and many others. The metaverse is certainly the second most well-known and universally recognized arm of the Web 3.0 ecosystem, just behind cryptocurrency (Park & Kim, 2022). The explosion of this interest is happening due to two reasons. First, the record-breaking sales of digital real estate. Second, when Mark Zuckerberg announced to change of the name of Facebook to Meta.

In the technological industry, every researcher is coming up with their own explanation of the metaverse. Meanwhile, other researchers believe that it will still take several years to enter the new era of the internet. Due to this conflict going on around, when renowned author and entrepreneur Jon Randoff vocalized a simple and systemic explanation of the metaverse, it instantly gained popularity among Web 3.0 enthusiasts (Entrepreneur, 2022). According to him, the Metaverse comprises seven fundamental layers that provide the systematic outline for its architecture and explain the stages of the metaverse economy including technological innovations, opportunities, and solutions to our current problems (Blockchain Council, 2022a). In other words, the seven-layer concept of the metaverse is a dedicated framework of output, structure, and ethical framework of an idealized decentralized metaverse. Jon Radoff has promoted these seven layers as a guiding map of the metaverse that is expected to benefit creators and compensate them fairly (Times of India, 2022).

The seven layers are experience, discovery, creator economy, spatial economy, decentralization, human interface, and infrastructure (Fig. 3.1).

Fig. 3.1   Seven layers of the metaverse (*Source* Far and Rad [2022])

3.2   Layer 1—Experience

Currently, the only way of experiencing a tiny fraction of the metaverse individuals is by logging on the platforms like Sandbox or Decentraland and exploring the digital open world or maybe buying and trading cryptocurrencies. However, it is expected that the full-fledged form of the metaverse will not be limited to the 3D world on the browser rather it will be a step ahead of the internet, i.e., immersive space connecting devices at the home, workplace, and at social events like concerts (Bale et al., 2022). For instance, products like MetaQube are focusing on enacting game thinking and game logic to the problems in training, EdTech, and industrial sectors.

As the physical space becomes dematerialized, all the physical constraints will be lifted and the metaverse will be able to provide an abundance of experiences that one is devoid of today. For instance, not able to secure a front-row ticket for a concert? Metaverse will provide the front-row experience for all the tickets. The immersive and real-time aspect of the metaverse may revolutionize a wide range of human activities, like gaming, eCommerce, social interactions, entertainment, and e-sports (Lee et al., 2021a).

In recent times of the COVID pandemic, we have seen how movie theatres have struggled to attract audiences and remain open. Most moviegoers now prefer to sit at their homes and browse in their comfort and safety. In March 2021, Pew Research Center conducted a study indicating that 31% of Americans were nearly always connected to the internet, while 48% logged on multiple times per day (Pew Research Center, 2022a) (Fig. 3.2).

According to Arun Maini, a prominent Tech YouTuber from England with 9 million subscribers, there is a noticeable trend in which individuals are moving away from physical items and towards virtual goods, as reflected by the amount of time spent on apps and games (ABC News, 2022). This shift from the tangible world to the digital realm is already taking place and can be observed by the declining success of films in movie theatres.

The Indian entertainment and media conglomerate, Shemaroo Entertainment has launched its Web3 campaign, and partnered with metaverse consulting and development firm Filmrare, on one of the leading metaverse platforms, Decentraland. This open and immersive cinema is named "Shemaroo Theatre" (Economic Times, 2000a). With the introduction

More than 8 in 10 U.S. Adults go online at least daily

% of U.S. Adults who say they go online

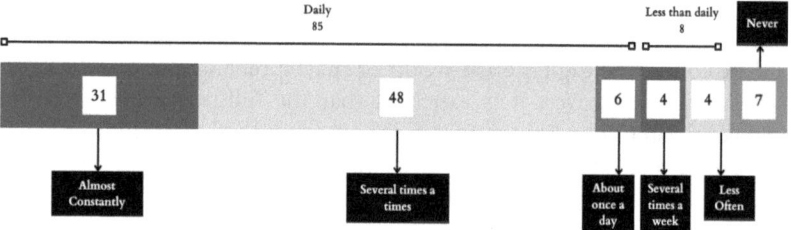

Fig. 3.2 Survey of online US adults on online activity (*Source* Pew Research Center [2022a])

of the metaverse, Shemaroo aims to popularize Bollywood on a global stage and present hits across eras through its large repository of content.

Shemaroo Theatre is expected to create an avenue to offer the "First cinema experience" with the best Bollywood entertainment content to the world. It has been created to give the same real-life movie theatre experience to its users. Virtual visitors will be able to explore the exciting plush lobby, box office counter, virtual trailer areas, and even including virtual popcorn and beverages counters for immersive storytelling (Binance, 2022).

3.3   Layer 2—Discovery

This section pertains to experiential revelations that occur through the ongoing exchange of information. This exchange introduces users to novel experiences through a process of "push and pull." The "pull" aspect refers to an inbound system where users actively seek out information and experiences. However, on the other hand, the push is more outbound that involves the process of informing users about the experience that the metaverse awaits them in the metaverse (Li et al., 2022). In other words, Inbound discovery takes place when people are actively seeking information related to the topic through apps, search engines, or rating sites. Meanwhile, outbound refers to the method of putting out push messages in the form of emails, advertisements, or notifications to people whether they want it or not. Inbound and outbound activities can take place in the following ways:

Inbound

- Real-time presence
- Search Engines
- Community-driven content
- Earned Media
- App Stores

Outbound

- Notifications
- Emails and social media
- Display advertising

As communities keep growing and evolving, individuals will get more opportunities to learn about newer products, services, and applications through personal recommendations and a curation-based approach. As the anxiety around personal data and advertisements increases, the metaverse will show more dependence on organic interactions of the users rather than data mining for providing targeted services (Golf-Papez et al., 2022). This will further help lower the marketing cost and make the process more streamlined and powerful for people on both sides of the aisle. This layer is more lucrative in the business.

Radoff argues that content driven by community is a crucial and cost-efficient way to discover new things. In the era of influencers, one can witness the rise of content creation which makes it simpler to swap, share, and trade across different metaverse situations (Medium, 2021). As a discovery, the content marketplace will replace the application market-place. A recent example can be seen in the form of NFTs. These digital assets were extensively used by big brands as a marketing tool.

One of the great examples of a brand that marketed its product through NFTs is the Norwegian Cruise Line. To celebrate the launch of their new Norwegian Prime Class, the brand collaborated with artist Manuel Di Rita for creating six NFT art pieces. Each piece has been put up for auction starting from USD 2500 and proceeds will be further donated to Teach for America. In the press release, it mentioned that it thought to promote its new product through NFTs as its cutting-edge technology and will eventually reflect how it has approached its products and services. By doing so, Norwegian was able to leverage the buzz about NFTs to create buzz and awareness towards its launch (Norwegian Cruise Line Holdings Limited, 2022).

Another major contributor to inbound discovery is real-time presence. At its core, the metaverse revolves around building interpersonal relationships through collective experiences. While community-generated content is significant, it's also important to take into account the experiences of others to uncover the potential of the metaverse. As an example, when someone logs into platforms like Spotify, Xbox, or PlayStation, they can view the activities of their friends, which provides insight into what others are doing.

Gaming systems have increased in-game involvement by using real-time presence (Marchand & Hennig-Thurau, 2013). Games are employed strategically to enhance the interactivity within them. Additionally, non-gaming platforms like Clubhouse have effectively utilized the strength and adaptability of real-time presence by allowing individuals to choose chat rooms based on the locations of their friends. In the metaverse, real-time presence will be crucial to improving user interactions, which, in turn, will increase comprehension of the virtual universe.

The metaverse has the potential to convert social structures into digital form and establish a decentralized identity system, thereby transferring power from a small number of dominant identities to social groups. It will assist in allowing a frictionless exchange of experiences and information.

3.4   Layer 3—Creator Economy

The creator economy refers to the abundance of design tools and applications used by content creators and developers for producing immersive experiences, digital resources, and other assets. As time goes on, more platforms are incorporating drag-and-drop functionality to streamline the process of content creation (Fang et al., 2021). Though it has never been simpler to become a developer, designer, or creator and it is expected that it will only become easier as Web3 becomes more ingrained in the culture and Web2 gets phased away over time. Its existence can already be noticed on various Metaverse platforms such as Sandbox which helps in making the process of producing digital assets exceedingly simpler and code-free (Fig. 3.3).

Jon Radoff suggests that by providing creators with tools, templates, and content marketplaces, development can shift from a bottom-up,

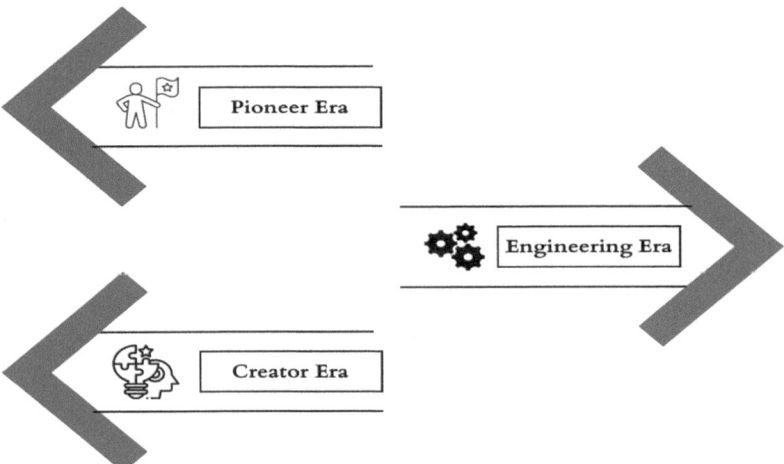

Fig. 3.3 Evolution of creator economy (*Source* Innovius Research [2022])

code-driven process to a top-down, creatively driven process. The creator economy has evolved in three phases which are the pioneer era, the engineering era, and the creator era.

Pioneer Era: This was the time when companies like Pixar and Amazon created their own technologies. In this era, to create any experience for a specific technology, people have to build everything from scratch due to the lack of availability of any tools to support it. The first web page was directly coded in HTML, individuals created their own shopping carts for e-commerce sites and game developers wrote code directly for graphics hardware for games (Innovius Research, 2022).

Engineering Era: This was when bottoms-up tools and middleware emerged to support overwhelmed engineering tools. Writing codes from scratch was often too slow and costly for people to meet the requirements, and the workflow was more complicated. The earliest tools available in the market tend to reduce the workload of engineers by providing SDKs and middleware to save time and avoid building everything from scratch. For instance, Ruby on Rails significantly reduces the time and effort of engineers in creating data-driven websites. In games, the availability of graphics libraries like OpenGL and DirectX helps programmers to create 3D graphics without understanding a lot of low-level coding.

Creator Era: This is the time when top-down tools emerge to support a much larger creators market and disrupt many businesses of the prior era. In this era, creators and coders do not want to slow down their speed due to coding bottlenecks, they want to enhance and utilize their skills to create unique aspects of the project (Prayitno, 2022). There has been rapid and exponential growth in the number of creators in this era who have access to different and a vast variety of tools, templates, and content markets that shift the development process from a bottom-up, code-centred approach to an upward-facing, aesthetically driven process.

Now, people can create 3D graphic experiences in games such as Unity and Unreal without even touching or learning any low-level API which uses the visual interface in its studio environment. Even one can create an e-commerce website in Shopify without knowing any coding (Dwivedi et al., 2022a).

In the era of Web 3.0, also known as the creator era, anyone will have the ability to become a creator on the web without dedicating hours to programming. The Creator Economy is characterized by a significant rise in the number of creators, who will be able to transform the metaverse into a lucrative opportunity by showcasing and selling commercial goods, NFTs, and in-real-life products (Wang et al., 2022a).

Content creators will be crucial in shaping this new world. On social media platforms, creators have achieved immense success and remain a significant growth driver in the virtual realm of the metaverse. According to Bojic (2022), it is anticipated that the metaverse will revolutionize the creator economy, turning it into a multibillion-dollar industry. This economy includes self-employed individuals who generate various digital content such as videos, images, webinars, e-books, artwork, blog posts, and other digital goods.

As the metaverse evolves, it's expected that this group of creators will seek to profit from it. They will have the ability to construct their own metaverse realms, where they can connect, socialize, and engage with their audience. With technological progress, creators will also be able to easily relocate their followers to the metaverse (Xu et al., 2022a). It should be emphasized that developing skills is crucial for creators to fully leverage the potential of the metaverse.

Creators will have the opportunity to monetize their presence in the metaverse by:

- Selling and displaying NFTs.

- Influencing buying behaviour by making new avatars by wearing virtual fashion apparel and accessories like clothes and trainers.
- Hot get-togethers and parties for their followers to strengthen relationships and boost sales (Bojic, 2022).
- Selling commercial goods, IRL, and NFTs products.
- Collaborating with brands and promoting them.

Emergen Research's study revealed that the metaverse was valued at USD 63.08 billion in 2021, and it's predicted to rise to USD 1607.12 billion by 2030, with a compound annual growth rate of 43.3% between 2021 and 2030 (Emergen Research, 2022). The surge in demand for products and applications utilizing virtual, augmented, and mixed reality is anticipated to be a crucial driving force behind the growth of the global market revenue (Emergen Research, 2022).

Hence, Metaverse will open up a new and promising market for the creator economy. Also, it is expected that the condition will be a win–win for all, that is, the user community, the creators, and the metaverse.

3.5   Layer 4—Spatial Computing

The term spatial computing refers to a tech solution that combines AR, VR, and MR for bringing the metaverse to life. As per Jon Radoff, spatial computing helps in controlling, manipulating, and exploring 3D spaces (Egliston & Carter, 2022). It facilitates the digitalization of objects using the cloud, enables sensors to react with motors, and digitalizes the physical world around us through spatial mapping. Spatial computing has already made lives easier through ride-handling apps and virtual assistants. By offering virtual fitting rooms, it has become possible to add some enjoyment to fashion and make shopping more convenient for customers. Currently, spatial computing is preparing to provide individuals with the ability to work, shop, and socialize as avatars in a three-dimensional digital world that resembles the real world.

A game that utilizes spatial computing allows players to engage with it while surrounded by their actual real-world environment. The game characters will recognize physical objects in the vicinity and can interact with them, such as sitting on a living room sofa. (Hawkins, 2022a). Spatial computing essentially enables real-time interaction with both the real and virtual worlds concurrently. Spatial computing includes:

- Virtual, augmented, or mixed-reality technologies
- Internet of Things devices (such as sensors in warehouses and on robots)
- Speech recognition (such as Amazon's Alexa and Apple's Siri) (Hawkins, 2022a)
- Gestural recognition

In spatial computing, the method of user interaction differs significantly from the conventional computer interface that most of us use, which involves typing or touching a fixed screen (Nalbant & Uyanik, 2021). In the future, spatial computing will utilize eye-controlled interactions, body or hand gestures, and voice controls to make hardware unobtrusive, resulting in a shift away from fixed computers and flat screens. The amalgamation of real and virtual worlds through mapping the layers of virtual information on the physical spaces that people use every day has become a very exciting field for marketing and technical people. According to a Harvard Business Review Study, shoppers who used AR spent more time browsing and viewing more products. Customers who used AR viewed 1.28 more products and were 19.8% more likely to make purchases compared to customers who did not use AR (Harvard Business Review, 2022b).

Recent examples of spatial computing are Microsoft's Hololens and Snapchat's Landmaker. The most widely used example is the face filters on Instagram or the massively popular 2016 game, Pokemon Go (Lee et al., 2021a). All of these were possible because of spatial computing. Also, VERSES' Wayfinder creates a 3D model of warehouses, which allows companies to work more efficiently by optimizing their inventory location and storage spaces. It even helps in evaluating the best routes for workers in real-time, cutting the travelling time between pickups. Further, this helps in making workers more efficient and picking more items.

Virtual 3D tours of any property are doable whether it is for sale, under construction, or still in the designing phase. Spatial mapping helps prospective clients in a detailed view irrespective of their location. This is also helpful for interior designers and contractors (Alpala et al., 2022). Also, Tesla uses a variety of spatial sensors to recognize the surrounding environment, objects, and people. Further, In Amazon's automated warehouses, robots have been deployed to pick and pack the products. Through the number of sensors, proximity calculations, and coordinates, they can safely navigate to the specific location in the warehouse and pick the product from the correct physical shelf, put it in a bin, and return it to its original place (Vox, 2019). All of this they can do while navigating safely through physical structures, humans, or other robots in warehouses.

Over time, spatial computing has become a significant category of technology, enabling individuals to manipulate and access 3D space for enhanced experiences. Specialized hardware and software are essential for the optimal functionality of spatial computing. The hardware aspect is more related to the "human interface" layer. The spatial computing layer is more related to the software aspect. Various aspects of the software layer that power the metaverse are listed below:

- Voice and gesture recognition.
- Human biometrics for identification purposes (Jiang et al., 2022).
- One necessary tool for spatial computing is a 3D engine that can display and animate geometry. Popular examples of such engines include Unity and Unreal Engine
- The Internet of Things (IoT) integrates data from various devices.
- Geospatial mapping and object recognition map and interpret both the virtual and real world. This is according to Jiang et al. (2022).
- Next-generation user interfaces support concurrent information streams and analysis.

3.6   Layer 5—Decentralization

Roundhill Investment predicts that the metaverse industry will experience a compounded growth rate of 13.1% in the coming years and will likely accelerate further as the metaverse network becomes more widely available (Roundhill Investments, 2021).

While major tech companies are generating hype about entering the metaverse, it is evident that they will have a crucial role in the evolution and progress of the metaverse (Dwivedi et al., 2022a). This poses an important concern: Will these companies pose the same privacy and data security challenges that the internet currently faces? For instance, Facebook's business model relies on possessing significant user data, which it uses to enable third parties to display personalized advertisements to users.

One of the defining characteristics of the metaverse is expected to be its lack of central authority and ownership, making it accessible and available to everyone. Decentralization, along with openness and distribution, is considered to be a crucial feature of the metaverse in an ideal scenario (Mircică, 2022). Alternatives increase when systems are interoperable and built inside competitive marketplaces which leads to an increase in growth and experimentation. Moreover, creators have control and ownership of their own data and products. But central ownership makes it impossible for normal users to determine who was privy to it and under what conditions. This may result in security breaches.

The decentralized and open nature of the metaverse raises concerns about privacy and data security issues. To address these concerns, blockchain technology is seen as a potential solution. Unlike structured tables in a centralized database, blockchain technology stores data in blocks that are linked in chronological order. Each block contains data by users, and these blocks are stored locally on user devices and synchronized with other blockchain data on peer devices using a consensus model. In blockchain technology, users are referred to as nodes. (Kogure et al., 2017) (Fig. 3.4).

Some of the characteristics of blockchain include:

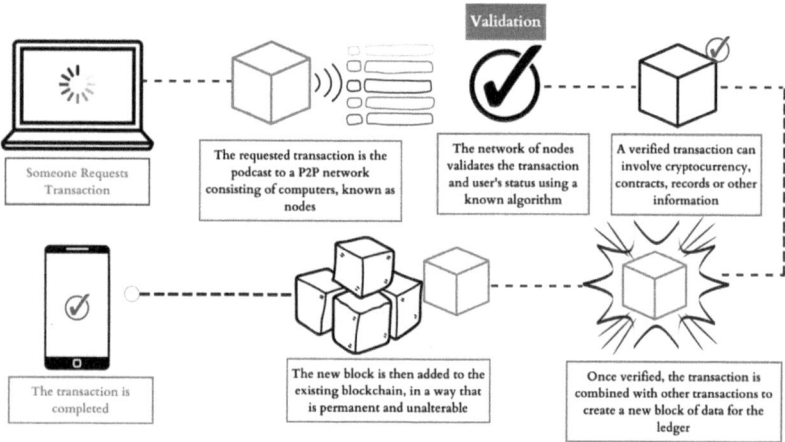

Fig. 3.4   How does Blockchain work? (*Source* PwC [2022])

Decentralization: The P2P network allows all users to manage the data stored in the blockchain in an equivalent and decentralized manner. The data is updated in a centralized manner.

Anonymity: All the transactions are addressed based on cryptographic algorithms, rather than personal identifications protecting the real identity of the user.

Immutability: To ensure data security and integrity, each block in the blockchain contains a hash value of the previous block, and this process continues in a chronological order as new blocks are added (Lee et al., 2021a). Therefore, if any modification has to be made in a block, all the subsequent blocks need to be regenerated and verified by all the users in the network to prevent security breaches.

Auditability: Each transaction is recorded with a timestamp. Blockchain ledgers are transparent and any changes made are documented, preserving integrity and trust. A complete traceability chain is formed with a timestamp, which could be used for auditing.

Autonomy: Data can be shared more freely and in a secure way through consensus algorithms and protocols without the interference of authoritative third parties (Lee et al., 2021a).

Several decentralized metaverse initiatives are incorporating blockchain technology to create a user-centric experience that is resistant to censorship and encourages interoperability. Decentraland is one such project that uses blockchain and is a decentralized virtual world powered by the Ethereum blockchain. It is operated by a Decentralized Autonomous Organization (DAO), which can modify its policies through a voting system (Consensys, 2021).

To fully leverage the capabilities of the metaverse, it will be essential to have a clear and open method of recording and managing interactions and transactions. Blockchain and cryptocurrency offer a possible solution to this issue, as they enable traceability and transparency (Mircică, 2022). Furthermore, NFTs can be used to avoid any conflicts over asset ownership in the metaverse. While the metaverse itself is impressive, crypto assets, blockchain, and NFTs will undoubtedly have a significant impact on the realization of its full potential.

3.7   Layer 6—Human Interface

To fully experience the metaverse, the hardware layer, or the devices that enable user interaction, is crucial. The human interface layer is a key component of the hardware layer, which encompasses the necessary technologies for users to interact with the metaverse through human–computer interaction, as noted by Lee (2022). By combining both spatial computing and human interface, gathering information about surroundings has been easy.

Now even AR experiences can be created by just looking around the physical world. A case in point is Amazon's latest augmented reality shopping tool, the Room Decorator, which enables users to visualize furniture in their own space. Though there have been many AR tools related to the decor in the past, Room Decorator allows users to add multiple products at the same time and see how a whole set of new products could fit in their physical space (Yahoo Finance, 2020).

This layer mainly includes VR headsets, smart glasses, and haptic technologies through which users can explore and navigate digital worlds. Further understanding of the topic and categorization of devices can be done based on AR and VR technologies.

AR Devices: The main characteristic of AR is the fact that it adds virtual elements into the real world. Because of this, users can directly interact through devices like smartphones, PCs, tablets, or smart glasses/ HDMs (Head-Mounted Displays). The latter are designed as small displays integrated within a pair of eyeglasses or helmet (Hawkins, 2022a). What distinguishes them is the fact that they superimpose virtual elements to the real vision of the user, similar to the Room Decorated tool introduced by Amazon.

VR Devices: VR has a world of its own. It is digital and is made by the combination of software and hardware devices in which user experiences are immersive and not just interactive. It is more dependent on immersive headsets and other devices such as sensors, haptic devices, gloves, and earphones to provide a truly multisensory experience (Radianti et al., 2020). With the help of haptic technology, one can control electronic devices mid-air without touching any buttons or the screen. Some of its experimental models also include a feature where the user can easily feel the shape and texture of the virtual object.

3.8 LAYER 7—INFRASTRUCTURE

The infrastructure layer helps in connecting the devices to a network for facilitating the content and experiences underpinning the metaverse. For instance, 5D technology is dramatically improving the bandwidth and reducing the network latency, paving the way towards a better metaverse experience, such as more responsive real-time virtual reality sessions with friends. In addition to network connectivity, innovations in the semiconductor and GPU industries can help in paving the way towards more realistic and fluid experiences (Ning et al., 2021). Furthermore, greater processing power can also help in unlocking new AI capabilities for paving the way for more realistic and immersive experiences from the software perspective. 5G networks will drastically increase the capacity while also reducing the competition and latency. 6G will improve speeds by an additional magnitude.

To achieve untethered functionality, miniaturization, and high performance in the next generation of smart glasses, mobile devices, and wearables, powerful and compact hardware is required. This includes semiconductors that are expected to soon reach 3nm processes and beyond, as well as microelectromechanical systems (MEMS) that enable tiny sensors. Additionally, long-lasting batteries that are compact will be crucial. These insights were reported by Yahoo Finance in 2020.

This layer is concerned with the technological foundation necessary to establish a complete and compatible metaverse. Five technology clusters that power the metaverse are:

1. Augmented Reality and the virtual world.
2. The hardware and software infrastructure necessary for the development of a functional and interconnected metaverse includes computing power and network, edge computing, spatial positioning algorithms, GPU servers, virtual scene fitting, and real-time network transmission.
3. To provide users with personalized experiences that adapt to their changing preferences over time, the metaverse requires display technologies such as MR, VR, AR, XR, and ER, in addition to immersive audio-visual capabilities. (Rehm et al., 2015).
4. Virtual gaming technology, including 3D game engines like Unreal Engine and Unity, is used to create various resources such as sounds, images, and animations.

5. The use of blockchain technology, along with decentralized value transfer, smart contracts, and settlement platforms, will ensure ownership of value and its circulation within the metaverse.

As far as the technical aspects are concerned, the metaverse stands on the foundation of integrated circuits, powerful computer communication components, advanced display systems, network equipment, missed reality equipment, precision freedom optical systems as well as sophisticated and high-resolution cameras (Tang et al., 2022).

Further, for the devices to work efficiently mentioned in the human interface layer, there is a requirement for tiny and powerful hardware. According to Radoff, it includes semiconductors that are approaching the 3nm process and beyond, compact long-lasting batteries, and micro-electromechanical systems (MEMS), that facilitate tiny sensors. The fundamentals of infrastructure lie within three main aspects (Park & Kim, 2022). These are computation, communication, blockchain, and storage.

Computation and Communication: Computation and communication are critical components of the metaverse. The metaverse is a multimedia system that demands substantial computational power, but it must also be available at all times and from any location, necessitating a reliable communication infrastructure to support it (Cai et al., 2022). Enhancing computation and communication are always topics of cutting-edge research, not only limited to the metaverse. As a result, it is crucial to focus on improving these aspects to enhance the user experience in the metaverse.

Blockchain and Storage: The expectation from the metaverse is the ability to connect everyone in the world so that it can produce and store enormous data around maps, roles, etc., in the large storage, which is another basic infrastructure (Gadekallu et al., 2022). Further, in order to guarantee fairness and decentralization, the blockchain should be introduced to support sustainable ecosystem operations in the metaverse (Duan et al., 2021). Smart contracts, introduced by advanced blockchain systems like Ethereum, have expanded the functionality of blockchain by enabling DApps to operate. This has broadened the range of applications for blockchain and made it possible for the metaverse to establish a decentralized social ecosystem.

The market metaverse map in Fig. 3.5 shows that various companies have already been working on the development and extensive usage of various layers of the metaverse. Well-known companies like Meta,

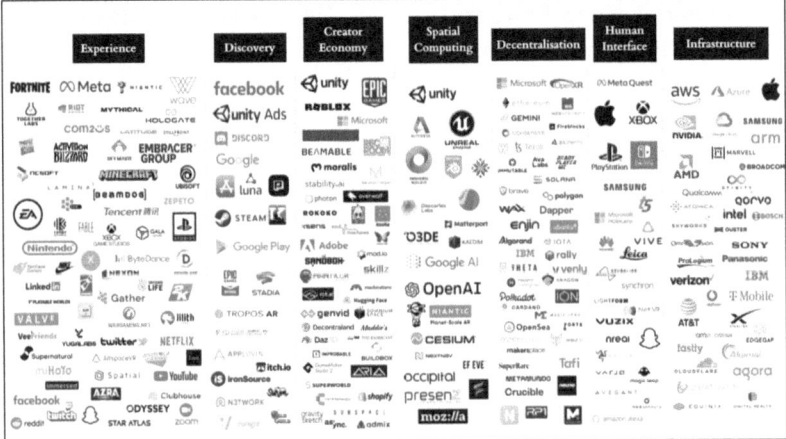

Fig. 3.5 Metaverse market map (*Source* Market place Fairness [2021])

Wave, Netflix, Clubhouse, YouTube, LinkedIn, etc., are working on the experience aspect of the metaverse. Google, Iron Source, Google play, Tuna, etc., are working at the discovery layer. Companies like Microsoft, Photon, Adobe, Shopify, Admix, etc., are more focused on the creator economy. Spatial computing aspects are focused on by the companies like Unity, Google AI, Matterport, OpenAI, etc. (Market place Fairness, 2021). Further, companies like Dapper, IBM, ION, Crucible, Solana, etc., are concentrating on the decentralization aspect of metaverse. The human interface layer has been pioneered by the companies such as Samsung, Microsoft Hololens, Apple, Avegant, Amazon Alexa, etc. Ultimately, companies like Intel, Nvidia, AWS, Bosch, Sony, etc., are building infrastructure aspects of the Metaverse. However, it is important to state that no one company has pioneered all the seven layers of the metaverse rather the concentration can be noticed more on one or two layers to ace the business game (Market place Fairness, 2021).

REFERENCES

ABC News. (2022). *How the metaverse could impact the world and the future of technology.* https://abcnews.go.com/Technology/metaverse-impact-world-future-technology/story?id=82519587. Accessed 6 Nov 2022.

Alpala, L. O., Quiroga-Parra, D. J., Torres, J. C., & Peluffo-Ordóñez, D. H. (2022). Smart factory using virtual reality and online multi-user: Towards a metaverse for experimental frameworks. *Applied Sciences, 12*(12), 6258.

Bale, A. S., Ghorpade, N., Hashim, M. F., Vaishnav, J., & Almaspoor, Z. (2022). A comprehensive study on metaverse and its impacts on humans. *Advances in Human-Computer Interaction.*

Binance. (2022). *Shemaroo launches "Shemaroo Theater" on decentraland.* https://www.binance.com/en/news/flash/7223148. Accessed 6 Nov 2022.

Blockchain Council. (2022a). *Decentraland metaverse: A complete guide.* https://www.blockchain-council.org/metaverse/decentraland-metaverse/. Accessed 22 Nov 2022.

Blockchain Council. (2022b). *Sandbox metaverse: An ultimate guide.* https://www.blockchain-council.org/metaverse/sandbox-metaverse/. Accessed 21 Nov 2022.

Blockchain Council. (2022c). *Understanding the seven layers of the metaverse technology.* https://www.blockchain-council.org/metaverse/seven-layers-of-the-metaverse-technology/. Accessed 6 Nov 2022.

Bojic, L. (2022). Metaverse through the prism of power and addiction: What will happen when the virtual world becomes more attractive than reality? *European Journal of Futures Research, 10*(1), 1–24.

Cai, Y., Llorca, J., Tulino, A. M., & Molisch, A. F. (2022). Compute-and data-intensive networks: The key to the metaverse. *arXiv preprint,* arXiv:2204.02001.

Consensys. (2021). *What is a DAO and how do they work?* https://consensys.net/blog/blockchain-explained/what-is-a-dao-and-how-do-they-work/. Accessed 6 Nov 2022.

Duan, H., Li, J., Fan, S., Lin, Z., Wu, X., & Cai, W. (2021). Metaverse for social good: A university campus prototype. In *Proceedings of the 29th ACM International Conference on Multimedia* (pp. 153–161).

Dwivedi, Y. K., Hughes, L., Baabdullah, A. M., Ribeiro-Navarrete, S., Giannakis, M., Al-Debei, M. M., ..., Wamba, S. F. (2022a). Metaverse beyond the hype: Multidisciplinary perspectives on emerging challenges, opportunities, and agenda for research, practice and policy. *International Journal of Information Management, 66,* 102542.

Dwivedi, Y. K., Hughes, L., Wang, Y., Alalwan, A. A., Ahn, S. J., Balakrishnan, J., ..., Wirtz, J. (2022b). Metaverse marketing: How the metaverse will shape the future of consumer research and practice. *Psychology & Marketing, 40*(4), 750–776.

Economic Times. (2022a). *Now Shemaroo offers movie theatre on metaverse.* https://economictimes.indiatimes.com/tech/technology/now-shemaroo-offers-movie-theatre-on-metaverse/articleshow/94538342.cms?from=mdr. Accessed 6 Nov 2022.

Economic Times. (2022b). *Rise of metverse in global Pharma industry*. https://health.economictimes.indiatimes.com/news/pharma/rise-of-metaverse-in-glo bal-pharma-industry/95250523. Accessed 22 Nov 2022.

Egliston, B., & Carter, M. (2022). 'The metaverse and how we'll build it': The political economy of Meta's Reality Labs. *New Media & Society*, https://doi.org/10.1177/14614448221119785.

Emergen Research. (2022). *Metaverse market, by component, by platform, by offering, by technology, by application, by end-use and by region forecast to 2030*. https://www.emergenresearch.com/industry-report/metaverse-market. Accessed 6 Nov 2022.

Entrepreneur. (2022). *Introducing 'touch' in the metaverse*. https://www.entrepreneur.com/en-in/technology/introducing-touch-in-the-metaverse/430606. Accessed 6 Nov 2022.

European Gaming. (2021). *Search data reveals absolutely no one understands the metaverse*. https://europeangaming.eu/portal/latest-news/2021/11/02/103027/search-data-reveals-absolutely-no-one-understands-the-metaverse/. Accessed 6 Nov 2022.

Fang, Z., Cai, L., & Wang, G. (2021). MetaHuman creator the starting point of the metaverse. In *2021 International Symposium on Computer Technology and Information Science (ISCTIS)* (pp. 154–157). IEEE.

Far, S. B., & Rad, A. I. (2022). Applying digital twins in metaverse: User interface, security and privacy challenges. *Journal of Metaverse, 2*(1), 8–16.

Gadekallu, T. R., Huynh-The, T., Wang, W., Yenduri, G., Ranaweera, P., Pham, Q. V., ..., Liyanage, M. (2022). Blockchain for the metaverse: A review. *arXiv preprint*, arXiv:2203.09738.

Golf-Papez, M., Heller, J., Hilken, T., Chylinski, M., de Ruyter, K., Keeling, D. I., & Mahr, D. (2022). Embracing falsity through the metaverse: The case of synthetic customer experiences. *Business Horizons, 65*(6), 739–749.

Harvard Business Review. (2022a). *How augmented reality can—And Can't—Help your brand*. https://hbr.org/2022/03/how-augmented-reality-can-and-cant-help-your-brand. Accessed 6 Nov 2022.

Harvard Business Review. (2022b). *How the metaverse could change work*. https://hbr.org/2022/04/how-the-metaverse-could-change-work. Accessed 20 Dec 2022.

Hawkins, M. (2022a). Metaverse live shopping analytics: Retail data measurement tools, computer vision and deep learning algorithms, and decision intelligence and modeling. *Journal of Self-Governance & Management Economics, 10*(2).

Hawkins, M. (2022b). Virtual employee training and skill development, workplace technologies, and deep learning computer vision algorithms in the immersive metaverse environment. *Psychosociological Issues in Human Resource Management, 10*(1), 106–120.

Innovius Research. (2022). *All about metaverse: 7 layers.* https://www.innovi usresearch.com/blog/market-report/7-layers-of-metaverse/. Accessed 6 Nov 2022.

Jiang, Y., Kang, J., Niyato, D., Ge, X., Xiong, Z., Miao, C., & Shen, X. (2022). Reliable distributed computing for metaverse: A hierarchical game-theoretic approach. *IEEE Transactions on Vehicular Technology, 72*(1), 1084–1100.

Kogure, J., Kamakura, K., Shima, T., & Kubo, T. (2017). Blockchain technology for next generation ICT. *Fujitsu Scientific & Technical Journal, 53*(5), 56–61.

Lee, J. (2022). A study on the intention and experience of using the metaverse. *Jahr: Europski časopis za bioetiku, 13*(1), 177–192.

Lee, L. H., Braud, T., Zhou, P., Wang, L., Xu, D., Lin, Z., ..., Hui, P. (2021a). From internet and extended reality to metaverse: Technology survey, ecosystem, and future directions. *arXiv e-prints*, arXiv-2110.

Lee, L. H., Braud, T., Zhou, P., Wang, L., Xu, D., Lin, Z., ..., Hui, P. (2021b). All one needs to know about metaverse: A complete survey on technological singularity, virtual ecosystem, and research agenda. *arXiv preprint*, arXiv: 2110.05352.

Lee, L. H., Lin, Z., Hu, R., Gong, Z., Kumar, A., Li, T., ..., Hui, P. (2021c). When creators meet the metaverse: A survey on computational arts. *arXiv preprint*, arXiv:2111.13486.

Li, H., Cui, C., & Jiang, S. (2022). Strategy for improving the football teaching quality by AI and metaverse-empowered in mobile internet environment. *Wireless Networks*, 1–10.

Marchand, A., & Hennig-Thurau, T. (2013). Value creation in the video game industry: Industry economics, consumer benefits, and research opportunities. *Journal of Interactive Marketing, 27*(3), 141–157.

Market Place Fairness. (2021). *How to invest in the metaverse in 2022.* https://www.marketplacefairness.org/cryptocurrency/how-to-invest-in-metaverse-2022/. Accessed 6 Nov 2022.

Medium. (2021). *Cryptovoxels: Do not miss the future of the metaverse.* https://medium.datadriveninvestor.com/cryptovoxels-this-is-the-future-of-the-met averse-4467326d4102. Accessed 20 Nov 2022.

Mircică, N. (2022). Immersive and engaging digital content, data visualization tools, and location analytics in a decentralized metaverse. *Linguistic & Philosophical Investigations, 21*, 89–104.

Nalbant, K. G., & Uyanik, Ş. (2021). Computer vision in the metaverse. *Journal of Metaverse, 1*(1), 9–12.

Ning, H., Wang, H., Lin, Y., Wang, W., Dhelim, S., Farha, F., ... & Daneshmand, M. (2021). A Survey on Metaverse: the State-of-the-art, Technologies, Applications, and Challenges. *arXiv preprint* arXiv:2111.09673.

Norwegian Cruise Line Holdings Limited. (2022). *Norwegian cruise line announces cruise industry's first NFT collection*. https://www.nclhltd.com/news-media/press-releases/detail/480/norwegian-cruise-line-announces-cru ise-industrys-first-nft. Accessed 6 Nov 2022.

Park, S. M., & Kim, Y. G. (2022). A metaverse: Taxonomy, components, applications, and open challenges. *IEEE Access, 10*, 4209–4251.

Pew Research Center. (2022a). *The metaverse in 2040*. https://www.pewres earch.org/internet/2022/06/30/the-metaverse-in-2040/. Accessed 3 Jan 2022.

Pew Research Center. (2022b). *About three-in-ten U.S. adults say they are 'almost constantly' online*. https://www.pewresearch.org/fact-tank/2021/03/26/about-three-in-ten-u-s-adults-say-they-are-almost-constantly-online/. Accessed 6 Nov 2022.

Prayitno, W. (2022). The "Metaverse" Symbol of Civilization Transfer in the Middle of Digital Economic Hegemony: Synthesis of Progressive Law of The Covid-19 Pandemic Era. *International Journal of Social Science Research, 4*(3), 14–32.

PwC. (2022). *Making sense of bitcoin, cryptocurrency and blockchain*. https://www.pwc.com/us/en/industries/financial-services/fintech/bitcoin-blockc hain-cryptocurrency.html. Accessed 6 Nov 2022.

Radianti, J., Majchrzak, T. A., Fromm, J., & Wohlgenannt, I. (2020). A systematic review of immersive virtual reality applications for higher educa tion: Design elements, lessons learned, and research agenda. *Computers & Education, 147*, 103778.

Rehm, S. V., Goel, L., & Crespi, M. (2015). The metaverse as mediator between technology, trends, and the digital transformation of society and business. *Journal for Virtual Worlds Research, 8*(2), 1–6.

Roundhill Investments. (2021). *Roundhill's intro to the metaverse*. https://www.roundhillinvestments.com/research/metaverse/intro-to-the-metaverse. Accessed 6 Nov 2022.

Tang, F., Chen, X., Zhao, M., & Kato, N. (2022). The roadmap of communica tion and networking in 6G for the metaverse. *IEEE Wireless Communications, 30*(4), 72–81.

The Times of India. (2022). *How the metaverse will change digital marketing*. https://timesofindia.indiatimes.com/blogs/voices/how-the-met averse-will-change-digital-marketing/. Accessed 21 Nov 2022.

Vox. (2019). *How robots are transforming Amazon warehouse jobs—For better and worse*. https://www.vox.com/recode/2019/12/11/20982652/robots-ama zon-warehouse-jobs-automation. https://www.roundhillinvestments.com/res earch/metaverse/intro-to-the-metaverse.

Wang, G., Badal, A., Jia, X., Maltz, J. S., Mueller, K., Myers, K. J., ..., Zeng, R. (2022a). Development of metaverse for intelligent healthcare. *Nature Machine Intelligence, 4*(11), 922–929.

Wang, H., Chen, D., & Deng, Q. (2022b). The formation, development and research prospect of educational metaverse. *Education Journal, 11*(5), 260–266.

Wang, M., Yu, H., Bell, Z., & Chu, X. (2022c). Constructing an edu-metaverse ecosystem: A new and innovative framework. *IEEE Transactions on Learning Technologies.*

Wang, X., Wang, J., Wu, C., Xu, S., & Ma, W. (2022d). Engineering brain: Metaverse for future engineering. *AI in Civil Engineering, 1*(1), 1–18.

Wang, Y., Su, Z., Zhang, N., Xing, R., Liu, D., Luan, T. H., & Shen, X. (2022e). A survey on metaverse: Fundamentals, security, and privacy. *IEEE Communications Surveys & Tutorials.*

Xu, M., Ng, W. C., Lim, W. Y. B., Kang, J., Xiong, Z., Niyato, D., ..., Miao, C. (2022a). A full dive into realizing the edge-enabled metaverse: Visions, enabling technologies, and challenges. *IEEE Communications Surveys & Tutorials, 25*(1), 656–700.

Xu, X., Zou, G., Chen, L., & Zhou, T. (2022b). Metaverse space ecological scene design based on multimedia digital technology. *Mobile Information Systems, 2022.*

Yahoo Finance. (2020). *Amazon's new room decorator tool lets you design a whole room with augmented reality furniture.* https://finance.yahoo.com/news/amazon-room-decorator-tool-lets-163700132.html. Accessed 6 Nov 2022.

CHAPTER 4

Metaverse Platforms and Use-Cases

Abstract Currently, metaverse is experiencing its adoption by popular tech companies such as Apple, Facebook, Microsoft, Accenture, and Disney. Its use-cases can also be noticed in industries such as gaming, education, travel, tourism, healthcare, banking, and a prominent functional area like human resource management. Though it is still at the developing stage, companies from these industries are trying their best to pursue the best of what is available in the metaverse world. This chapter discusses the adoption of metaverse and its use-cases in companies and popular industries in the world.

Keywords Apple · Facebook · Microsoft · Disney · Accenture · Gaming · Healthcare · Education · Banking · HR management · Social media

4.1 Major Platforms and Tools

In 2022, the search for a metaverse platform has been fueled by the growing interest in the metaverse's potential. Major technology companies like Microsoft, Facebook, and Nvidia are working on developing their own metaverse solutions. Several metaverse tools and platforms have already gained popularity in the market, and they will be discussed in detail in the upcoming sections.

R. Gupta and S. K. Pal, *Introduction to Metaverse*,
https://doi.org/10.1007/978-981-99-7397-2_4

4.1.1   Decentraland

Decentraland is considered a forerunner in the field of metaverse technology, offering a platform for trading, creating, monetizing, and exploring a virtual world. It provides users with the opportunity to express their creativity by designing artwork, challenges, scenes, and a variety of other virtual experiences (Blockchain Council, 2022a). Decentraland is a 3D virtual platform built on Ethereum that offers the buying and selling of virtual land. It is the first fully decentralized virtual space that provides users with a distinctive technological experience. In 2020, the platform was made accessible to the public (Decentraland, 2022).

The Decentraland metaverse offers a virtual world that replicates everything available in the physical world, including theatres, digital shops, skyscrapers, malls, and rapid transportation. As technology continues to advance, more people are becoming interested in the metaverse (Forbes,2022a). JP Morgan, the largest bank in the USA, has gone so far as to establish a lounge in Decentraland called Onyx Lounge, making it the first financial institution to enter the virtual world of the Decentraland Metaverse. In November 2021, the Metaverse Group, an NFT-based real estate organization, purchased a plot in Decentraland for approximately 18.15 crore rupees (approx. 2.2 million USD), according to The Hindu.

4.1.2   Sandbox

The Sandbox is a gaming metaverse that utilizes the Ethereum blockchain and NFTs. It allows users who may not have the technical expertise to create, sell, and earn money from their virtual reality NFTs. The Sandbox offers a range of integrated products that work together to provide users with a comprehensive gaming experience, enabling them to create their own user-generated content (McKinsey, 2022). The three integrated products are:

VoxEdit: The Sandbox enables users to produce voxel models, which are 3D pixels that are the fundamental building blocks of the Sandbox game. It is a package for creating NFTs on PC or Mac, which allows users to design and animate 3D objects, such as humans, animals, and vehicles.

Marketplace: In the Sandbox, users can upload, publish, and sell the voxel models they create using VoxEdit through the marketplace function. The models are first stored in a decentralized network called IPFS, and the users can register their ownership on the blockchain as proof (Blockchain Council, 2022a). After this process, the creations are transformed into assets that can be sold by the users through making a selling offer on the marketplace. Subsequently, interested buyers can make their purchases.

Game Maker: This feature allows anyone to develop their own 3D games at no cost. Even creators who lack coding skills can design impressive 3D games independently within a matter of minutes thanks to the user-friendly visual scripting tools.

Sandbox enjoys 39,000 daily users with 4.1 million total wallets and 1.6 million hours played (Blockworks, 2022).

4.1.3   Illuvium

Illuvium is an upcoming Ethereum blockchain-based fantasy role-playing game developed by a Decentralized Autonomous Organization (DAO) called Illuvium DAO. In January 2021, the game was first announced as an auto battler game and decentralized NFT collection, built on the Ethereum network (Blockworks, 2022).

Illuvium enables players to immerse themselves in a 3D open world where they can capture deity-like creatures called Illuvial. The Illuvials obtained by the players are represented as NFTs on Ethereum's blockchain and have real-world value. Players can battle with other players using their Illuvials and ether (ETH).

Illuvium also offers a unique feature in the form of a metaverse timeline, which adds to the immersive storytelling experience and game design. Its captivating gameplay has attracted not only gamers but also corporate players, making it one of the most popular platforms in this space.

Some of the notable companies that have shown interest in Illuvium include Influx, Immutable X, Blockchain game alliance, Bitcoin, and Lotus Capital (101 Blockchain, 2022).

4.1.4   Axie Infinity

Axie Infinity is a popular metaverse platform that offers a glimpse into the future of the metaverse. It is a gaming platform featuring magical creatures called Axies, which can be used for various tasks such as building, expanding, and defending the player's universe. Axies can also be used in battles and wars with other universes within the game.

Although it involves completing quests to advance to higher levels, Axie Infinity is more than just a typical online multiplayer game. As a metaverse platform, it stands out for its gaming applications that require players to employ new strategies and tactics to compete against each other. Additionally, the game incorporates blockchain technology to create a unique and complete economy within the game universe for the creatures to utilize.

In 2021, Axie Infinity was valued at $3 billion (Blockchain works, 2021). Also, with almost half a million daily active users and an estimated 60% of those coming from the Philippines, Axie Infinity is exploding (Coin Desk, 2021).

Apart from the Philippines, Axie Infinity is popular in Venezuela, the USA, Indonesia, Thailand, and Malaysia. It combines NFTs with a play-to-earn model, which makes it a perfect candidate for the major metaverse project. Furthermore, it has tangible similarities to the giant gaming behemoths of Pokemon and Tamagotchi (Analytics Insight, 2022a). Despite the disappointing start in 2022, in which the entire cryptocurrency market experienced a dip, Axie Infinity remains almost 5000% higher than it was a year ago showing incredible year-on-year growth (Coin Telegraph, 2022).

4.1.5   Cryptovoxels

It was rebranded as Voxels in May 2022, the name of Cryptovoxel in the most popular metaverse platform indicates how the metaverse is evolving. Built on the Ethereum blockchain, this metaverse platform is widely popular for providing a virtual gaming world that can be experienced through VR devices or PCs. Its distinctiveness lies in its regular organization of user events (Medium, 2021). This platform allows users to create virtual real estate and sell their properties. Users can generate empty plots of land for sale, as well as pre-built structures like art galleries, collaborative workspaces, harbours, and streets (Medium, 2021).

In addition to providing virtual real estate, Cryptovoxels has a large marketplace where users can buy and sell digital collectibles, including NFTs. The platform is regarded as one of the best metaverse platforms for supporting various types of NFTs, such as gaming props, wearable items, artwork, and apparel.

The virtual world in Cryptovoxels includes real-life infrastructure such as lands, roads, and buildings, among other details. The possibilities of creating anything on their own land are endless. Users can easily build, sell, or rent their land parcels (Tech Story, 2022). They also have the facilities to remove blocks, embedded audio, video, and images on their land parcels. Since it is a web-based metaverse platform, it can be accessed through any browser, such as Opera, Google Chrome, Safari, etc.

Cryptovoxels create a virtual world that operates similarly to the real world, allowing people to connect with others from different locations and time zones. For instance, if a planned concert is cancelled due to friends being unable to attend, users can still experience a similar concert together within the virtual world of Cryptovoxels. As per the data from MetaCat, Cryptovoxels rank fifth in terms of parcel sales volume with USD 38 Million (Meta Cat, 2022). The platform is regularly ranked just behind Decentraland and Sandbox in terms of user base and popularity (Globe Newswire, 2022).

4.1.6   Roblox

Roblox is known as a space for kids to play create, play, and engage with others. The platform gained popularity at the height of the pandemic as parents showed higher reliance on it to help their children to continue socializing in difficult times. Roblox hosts virtual experiences and games which include selling avatar customization and NFTs to earn "Robux," the digital currency of the platform. Roblox is an up-and-coming platform to keep an eye on, as it has the potential to help companies reach important demographics that they often overlook while converting prospective customers (Rospigliosi, 2022a). Using Roblox, corporations such as Nike and Hyundai have created areas for children to play and create a lasting impression of their brand.

For another instance, Hyundai Mobility Adventure and Hyundai's virtual experience allow future drivers to familiarize themselves with the vehicles and technologies of Hyundai through their five virtual theme parks. This helps in putting the steering wheels into the hands of the potential drivers who will recognize their logo and it with their positive childhood experience (Duan et al., 2021).

Roblox has recently partnered with the NFL to launch a metaverse game ahead of the 2022 Super Bowl. With millions of users worldwide, Roblox has a strong potential to drive the future of the metaverse. One of the most notable features of Roblox is its vision of bringing all experiences together under one roof, in addition to its diverse VR experiences.

4.1.7   Metahero

Metahero is another blockchain-based project that blends crypto with 3D scanning technology for creating accurate, life-like replicas of the world from people, and animals to cars in the metaverse. Instead of having video games like avatars, people can have avatars that look just like their real selves (Phemex, 2022).

Built on BSC blockchain technology, Metahero is optimistic that this technology can be applied in a wide range of industries such as gaming, fashion, medicine, and social media. It has also started a mission of accelerating the mass adoption of crypto by creating a larger database of people and canned objects.

Metahero can help by providing a platform where people can try out clothes on the online store before deciding whether they would buy them or not. It helps in making sure that people are buying is a good fit for them and reduces the possibilities of product returns. In the medical field, Metahero can be used while visiting a therapist. Rather than connecting through video streaming platforms, one can easily go for a live experience by visiting a doctor in the metaverse. The doctor can also use a virtual replica of the entire office that can provide a seamless metaverse experience to the patients.

Metahero is a total supply of 10 billion HERO crypto coins. As per the whitepaper of Metahero, 10% of the Metahero coin supply will be sold privately, another 10% will be sold publicly, and 20% of the HERO token supply will be locked in the liquidity pool (Metahero, 2022) (Table 4.1).

The sources of revenue in Metahero are NFTs generation fees, NFTs rolling royalties, 3D chamber scanning fees (including % of franchised meta scanner revenues), Metahero NFT marketplace transaction fees and partner licences for 3D chamber franchise. The platform is also trying to reach out to artists, galleries, and museums to use our technology to digitize and immortalize art collections and to earn extra revenue.

Table 4.1 Metahero's 5 year revenue projections as per paper released in 2022

	2022	*2023*	*2024*	*2025*	*2026*
3D Scanners	12	24	50	100	200
# of Scans	788,400	1,576,800	3,285,000	6,570,000	13,140,000
Metascanner Revenue ($)	157,680,000	315,360,000	657,000,000	1,314,000,000	2,628,000,000
NFT Marketplace Revenue ($)	100,000,000	250,000,000	500,000,000	1,000,000,000	2,000,000,000

Source Metahero (2022)

4.2   GLOBAL OUTLOOK

4.2.1   Apple

Tech giants are seeing huge opportunities in Metaverse. The iPhone and Mac maker has famously adopted a unique approach to software and hardware design. It is even incorporating that "Think different" ethos in its marketing. Reports suggest Apple will begin its entry into the metaverse with pricey headsets that bridge the gap between the virtual environment and digitally enhanced real-world use. It is expected that these new AR/VR headsets could hit the market in 2023 (Bloomberg, 2022).

Much of the metaverse in Apple is still in the development phase. The company is refining its implementation as much as possible with the focused release of new hardware. Further, Apple also enjoys a mature set of software for developers who want to incorporate AR in the applications for Apple devices. It includes ARKit and RealityKit development libraries, along with other tools such as Reality Converter and Reality Composer (Inc, 2022).

During an interview, Tim Cook disclosed that their yearly budget for research and development is approximately $25 billion (Meta Madrill, 2022). In addition, Apple's CEO has declared that the Apple store features over 14,000 apps for users, utilizing the capabilities of the Apple ARKIT development kit to provide easy access to various metaverse platforms (Forbes, 2022a). Many users are still waiting for official news to come from the CEO on the development and adoption of the Metaverse in Apple.

4.2.2   Facebook/Meta

Mark Zuckerberg, CEO of Facebook, launched Meta in 2021, which consolidates all of the company's applications and technologies under one new brand. Meta's goal is to make the metaverse a reality and aid people in connecting, finding communities, and growing their businesses. The term "metaverse" gained popularity after Facebook rebranded itself as "Meta." Zuckerberg views the metaverse as a successor to the mobile internet, and the company has made a substantial investment in this promising concept, employing over 10,000 personnel and acquiring AR and VR companies, while also providing significant financial support. Meta is engaging the services of some of the most talented engineering experts worldwide to ensure the success of the project (Xu et al., 2022a).

Though the timeline of this development is still unclear, Meta is determined to play a major role in building and shaping this new realm. In this area, Meta in 2021, launched VR meeting software Horizon Workrooms allowing users to conduct online VR meetings as "digital people" (Zuckerberg & King, 2021). For more than seven years, Meta has been quietly working on one of its most ambitious projects, a haptic glove that reproduces sensations like grasping an object or running your hand along a surface (The Verge, 2021). While Meta was not letting the glove out of its Reality Labs research division for many years, the company recently showed it off in 2021, alongside its other wearable tech, as a future of AR and VR interactions.

In an interview with Mark Zuckerberg, he revealed that the metaverse is not limited to virtual reality and can be accessed through various computing platforms, including virtual and augmented reality, personal computers, mobile devices, and gaming consoles.

Since there is still a lot to discover and implement, Meta would require continuous investment in the product as well as tech talent to grow the technology across the business. The company has already announced an investment of USD 50 million for collaborating with the government, other industry partners, academic institutions, civil rights groups, and non-profits to build this technology responsibly (News 18, 2021).

4.2.3 Microsoft

Microsoft has its separate vision for the buzzword of Metaverse. Microsoft, the owner of Xbox, has been constantly investing in games for the longest time. Moreover, Microsoft has revealed its plan to acquire Activision, the owner of Candy Crush and Call of Duty (PC Mag, 2022). The agreement between Microsoft and Activision indicates that individuals will spend an increasing amount of time in the digital area, and gaming will play a crucial role in metaverse-enabled virtual worlds.

The CEO of Microsoft, Satya Nadella, considers that if metaverse is associated with the engaging environment then gaming offers such experiences through titles such as Microsoft owned, Minecraft, and other game-making studios such as Mojang. According to the research conducted by Newzoo, gaming is expected to become a $218.7 billion market by 2024. The inclusion of Activision's games and personnel will assist the company in competing with gaming platforms such as Sony's PlayStation and Meta's Oculus (Newzoo, 2020).

Microsoft Mesh is a mixed-reality platform that enhances the Teams experience, enabling remote participants to join virtual meetings and collaborate on shared content using 3D avatars. Mesh has broadened the range of features in Teams such as Presenter mode and Together mode to enhance the collaborative and immersive aspects of hybrid and remote meetings (Zhang et al., 2018).

Microsoft's Mesh for Teams is accessible on a variety of devices, including traditional smartphones, laptops, and mixed-reality headsets. Additionally, it serves as a gateway to the Metaverse, a virtual world filled with digital representations of real-world places, individuals, and objects. As the world's second-largest public cloud server, Microsoft has expertise in edge computing and artificial intelligence (Deloitte, 2021a).

Additionally, AR headsets and the platform of Microsoft, Hololens, are already performing impressively well in the market. The platform is proving beneficial for manufacturing workers at companies like BMW, Daimler, and Ford. It enables them to learn faster, identify and correct errors, and enhance the overall quality of the manufacturing process (Coin Telegraph, 2021).

Microsoft plans to employ Metaverse applications in a variety of fields, including education, business, entertainment, and training. These real-world needs can be transformed and improved through the use of avatars, VR, and AR (Table 4.2).

Table 4.2 Microsoft vs. Meta comparison towards the metaverse technology

	▦ Microsoft	∞ Meta
Vision	Connecting the physical and digital worlds to allow for shared experience	Internet-based social media relationships
Focus	Enterprise offerings	Social element
Approach	Hierarchical Approach	Bottom-up approach
Early Concept	Mesh for Teams	Horizon Home
Innovation Perspective	Businesses may create their own virtual space using Teams	Develop and control its own metaverse

Source CoinTelegraph (2021)

4.2.4 Disney

Disney, an iconic entertainment brand renowned for creating some of the most beloved children's stories, is a perfect fit for the metaverse concept, which involves immersive online environments that offer endless potential for storytelling and magic. With properties like Marvel, Star Wars, and a vast library of animated films, Disney's IP would be a natural fit for any shared digital universe that revolves around its characters, making it a potentially huge source of revenue for the company (Forbes, 2022a).

Disney CEO Bob Chapek has recently appointed Mike White who has taken charge of "next-generation storytelling." In Disney's case, the metaverse is a new concept of entertainment supplied by VR devices. By utilizing technology it is aiming towards bridging the fictional world with reality. Disney has demonstrated a keen interest in the metaverse, as evidenced by the recent grant of a patent by the United States Patent and Trademark Office in 2022 (Verdict, 2022).

In early 2022, Disney was granted a new patent by the United States Patent and Trademark Office, detailing their strategy to develop a ride in a physical theme park that will incorporate a 3D virtual environment utilizing simultaneous localization and mapping (SLAM) technology. This method will allow visitors to the park to move around the actual world while also generating 3D projection imagery of their environment (Nasdaq, 2022). It will allow guests to see holograms, projected onto the 3D objects of their favourite Disney stars, information, or artworks further creating a highly immersive yet personalized experience.

Suppose the technology outlined in the patent is developed, Disney could potentially eliminate the expense of hiring actors to portray its characters at its theme parks. In addition to being cost-effective for the company over time, the technology could offer park guests a more authentic experience in which they can interact directly with the characters (Business to Community, 2022). Putting all the aspects together, these factors will ensure that Disney is well-placed to play a leading role in the development of the metaverse.

4.2.5 Accenture

Accenture believes that the metaverse is inarguably the most innovative and disruptive technology to debut in the market. The development of the internet has progressed to a point where users can do more than just browse online, as they can now take part in or live in a persistent shared experience that covers everything from the real world to the fully virtual and everything in between (Accenture, 2022a). Simply put, Accenture believes that the metaverse is a next-generation web that brings the physical and digital worlds closer.

Recently, in February 2022, Accenture created a living museum in the metaverse. Enabled by virtual reality headsets, participants were able to walk around the museum and explore the photos of black icons and pivotal moments in the civil rights movement (The Business Journals, 2022). Apart from the benefits that the museum portrayed to the visitors, it also gave a taste of the company's commitment to the visitors and demonstrated how the metaverse a used to connect with people in newer ways.

As per Bloomberg Intelligence, the metaverse landscape is expected to become a market of $800 billion by 2024 (Bloomberg, 2021). Accenture Interactive and more than 40 acquisitions from the past decade under a single umbrella to fortify synergies in product innovation, marketing commerce, and experience design. Accenture is distributing 60,000 virtual reality headsets to its staff worldwide to facilitate onboarding, learning, and collaborative immersive virtual reality environments (XR Today, 2021a). Within the next few years, Accenture anticipates that 80% of the workload will be hosted in the cloud (Accenture, 2021).

4.3   Industrial Use-Cases and Business Implications

4.3.1   Gaming

Metaverse has taken VR gaming to incredibly newer heights. Even the early immersive games in the 2D format, such as Minecraft and Second Life incorporated metaverse elements much earlier, such as world-building, 3D avatars, and observation as gameplay (Leeway Hertz, 2022a). Meta and Epic Games are currently working on creating a linked world of virtual realities with gaming at its core.

According to a recent survey by XR Today, 59% of industry experts predict that VR investments will mainly focus on gaming in the next few years, with 64% believing that gaming has the highest potential for VR among all other use-cases. Currently, VR gameplay is available as a standalone application that users can install on desktops, mobile phones, or VR gear for an immersive 360-degree experience of traditional video games. The only difference is that in VR, the game world is presented in a 3D format, giving users a realistic perception of the game environment that they can almost touch.

Some of the characteristics of gaming in the metaverse will include, Games-as-platform, social gaming, play-to-earn, portable games assets, and mixed-reality experiences. Companies like Sandbox, Decentraland, Epic Games, and Meta have become early movers in shaping the future of gaming in Metaverse (XR Today, 2021a). However, there are still certain challenges that can affect its adoption growth such as data security, incorporating NFTs, infrastructure breaches, managing user data, and controlling child-appropriate services and controls (Fig. 4.1).

The current industry report forecasts that the metaverse market will grow to $1,527.55 billion by 2029, with a compound annual growth rate (CAGR) of 47.6%, up from $100.27 billion in 2022. This growth is mainly driven by the increased adoption of online video gaming using AR/VR technologies.

4.3.2   Education and Learning

The education sector has evolved significantly since the advent of e-learning during the PC internet boom in the late 1990s. The emergence of social media and mobile computing in the second wave further

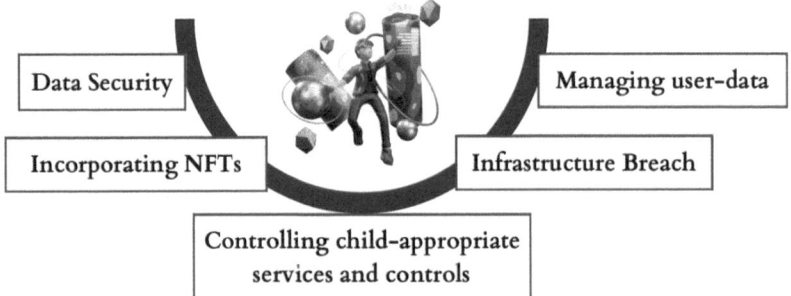

Data Security

Managing user-data

Incorporating NFTs

Infrastructure Breach

Controlling child-appropriate
services and controls

Fig. 4.1 Challenges to be addressed in Gaming Applications of Metaverse (*Source* Appinventiv [2022])

increased the popularity of video-based learning. Industry experts are now predicting that the third era of computing will introduce a new paradigm where the metaverse—digital 3D spaces where individuals can interact with life-like avatars—will replace PCs and phones. Unlike scheduled Zoom calls that disappear once completed, the metaverse is always available, providing opportunities for continuous social interactions with peers. This shift has far-reaching implications for capacity building and learning.

Meta has recently made an investment of approximately $150 million in Meta Immersive Learning, a move that is aimed at providing greater access to learning through the use of technology (Forbes India, 2022). The concept of utilizing the metaverse for educational purposes might appear to be a futuristic one; however, Indian start-ups have already begun investigating its potential. 21k school, an online-only school located in Bengaluru, recently revealed plans to incorporate Web3 technology in order to foster students' creativity. The school intends to introduce metaverse and blockchain technologies, such as NFTs, as tools to assist students in their creation and learning processes (Live Mint, 2022).

Invact Metaversity is a start-up that plans to launch a learning platform in the Metaverse to enhance the job readiness of students and provide an economical university-like experience within the Metaverse (Business Today, 2022). The metaverse learning experience is characterized by storytelling and visualization, offering a much-needed alternative to the monotonous Zoom experience during the Covid-19 pandemic.

By utilizing VR technology, learners can immerse themselves in entirely different worlds or take on the perspective of another person.

4.3.3   Healthcare and Pharmaceuticals

The precise impact of the metaverse on the life sciences sector is yet to be determined, but considerable research and development are in progress to leverage the metaverse and its related technologies for various life science purposes (Thomason, 2021). The integration of AI, AR, and VR in the metaverse has the potential to significantly transform the drug development process, offering pharmaceutical companies faster time-to-market and improved learning and development experiences.

MRFR has predicted that the metaverse will create new opportunities for providing affordable treatments and improving patient outcomes. The global healthcare industry is expected to grow at a CAGR of 48.3% during 2024–2030 and reach $5373 million by 2030 (Economic Times, 2022a).

As virtual reality and the metaverse gain more attention in the healthcare industry, major hospital chains and healthcare companies are starting to adopt these technologies to enhance patient care. This trend is expected to continue and gain even more momentum in the coming years (End Points News, 2022). Pharma companies invest millions of dollars in research and development, and the metaverse introduces amazing possibilities for using digital twins to starkly reduce the time and cost required for R&D.

To digitalize retail pharmacies in the virtual world, a more efficient supply chain could lead to quicker drug deliveries right to the doorsteps of customers. As the industry advances its manufacturing processes, demand for training resources to achieve desired business outcomes is expected to increase significantly (Chengoden et al., 2022). VR will not only allow faster employee training but will also endure more effective step-by-step experiential training enduring reduced training costs for Pharma companies in comparison to the traditional methodologies.

LifeNome, Digbi Health, XRHealth, Mindset Medical, iQ3 Connect, and NeuroTrainer have been named as finalists for the global challenge programme—Mission for the Future, in the healthcare and metaverse categories. The companies will work with LG NOVA on new business opportunities and compete for investments to create effective solutions, alongside other investors within the LG NOVA ecosystem (Inc 42, 2022).

4.3.4 Travel and Tourism

The metaverse is the next step to define the growing moment of the travel industry. The metaverse in travel and tourism is not limited to branded exploration and fun avatars. It has the capability for generating breakthrough value for businesses and stakeholders, including newer revenue streams. Over half of travel executives, or 53%, believe that the metaverse will bring favourable results to their organizations, and a quarter of them, or 25%, expect it to have a significant or transformative effect (Accenture, 2022a).

The metaverse can offer fresh methods for customer engagement, like integrating 3D games on cruise ships and providing additional services such as staying connected with guests in between trips (Revfine, 2022). The metaverse offers virtual spaces for customers to explore different room configurations and ship upgrades, providing them with novel ways to engage with the brand.

Hotels can benefit from the metaverse by simplifying guest experiences through a single platform for guests to access various services, ranging from entertainment purchases to logistical needs, such as booking theatre tickets and upgrading rooms or arranging airport transfers, through a virtual concierge (The Conversation, 2022).

Companies like Emirates, Qatar Airways, Qantas, Lufthansa, and Singapore Airlines are investing millions of dollars in metaverse experiences, as the Return on Investment associated with it is huge (Travel Dine, 2022). Private jet charter companies like Star Jets International are working with animators from Pixar, Disney, and 21st Century Fox to create metaverse assets for luxury travellers.

4.3.5 Banking and Finance

Currently, the banking industry has entered the fourth phase of evolution, with cryptos and NFTs coming to the forefront. However, there are certain banking institutions that have already entered the fifth phase of evolution, that is, the metaverse stage. JPMorgan Chase has opened a lounge and office in the metaverse, making it the first bank to do so. The Decentraland-based luxurious space was launched simultaneously with the publication of a paper by the bank's blockchain unit, Onyx. The paper predicts that the metaverse will impact every sector in some form in the near future (Fintech Magazine, 2022).

In March 2022, American Express applied for trademarks related to a virtual marketplace and cryptocurrency services in the Metaverse. The trademarks cover various real-world services, including payment services, banking, and fraud detection, as well as travel, entertainment, and concierge services that the company plans to offer to its virtual clients (Business Insider, 2022).

Quontic, an adaptive digital bank, provides banking services in its virtual offices, including a digital outpost with an interactive ATM, teller, posters, and guides to its website. The bank's debit card, launched in April, provides Bitcoin checking reward accounts, along with high-interest checking accounts and cash. By using the ATM, visitors can open a bank vault, enter a meta pool party, and receive free NFTs (Zeb Pay, 2022). Additionally, a digital bank called Congi acquired a selection of NFTs from Bored Ape Yacht Club and is developing a Bored Ape debit card for a new series of Web 3 activities for its clients. Similarly, HSBC announced its intention to open an office in the metaverse on Sandbox, while Siam Commercial Bank declared plans in March to launch a virtual headquarters on Sandbox (Wunderman Thompson, 2022).

4.3.6   Human Resource Management

To eliminate the boundaries between the physical and digital worlds, Meta and Microsoft are developing an all-encompassing digital realm that employs technologies like AR and holograms. Lenovo conducted a survey revealing that 44% of professionals are willing to work in the Metaverse, as they anticipate that it could improve work performance (Digite, 2021). Apart from this, a new survey was conducted among 1500 US bosses about working in the still nebulous concept of Metaverse.

According to the results of a recent survey of 1500 US managers, employees were less enthusiastic about adopting the metaverse in the workplace than their bosses were. The survey found that workers, in general, were more anxious and distrustful of the concept of the metaverse than their employers (Gizmodo, 2022). Larger organizations, with over 500 employees, were also more inclined to believe that the metaverse could have a greater positive influence on employee stress and work-life balance.

In the ExpressVPN survey, it was revealed that although 55% of respondents were aware of their employer monitoring them or their work, 73% of employers admitted to such surveillance. Among the surveyed workers, 63% expressed apprehension about their employer gathering their data in a metaverse office setting, including their current location and on-screen actions (Express VPN, 2022).

Meta's Horizon Workrooms and Microsoft's Mesh utilize virtual reality (VR) technology to enable teams to collaborate in a shared virtual space, regardless of their physical locations (Facebook, 2021). The new three areas of focus of metaverse in the workplace bring new opportunities to build a more equitable workplace, bringing change in the hiring process through the adoption of virtual recruitment fair at metaverse platform and setting up productive and collaborative workplace (HR TechX, 2022).

4.3.7   Social Media and Entertainment

In the metaverse era, social media platforms are expected to focus more on delivering engaging and interactive experiences that can activate multiple senses, as opposed to simply facilitating connections among friends through 2D webpages.

Xone is a Web3 social media platform that utilizes blockchain and NFTs to provide users with AR-based features, allowing them to build and share virtual worlds. Xones are the two types of zones where users create and interact: Personal Xones which serve as profile pages, and Community Xones which can be used to host events or other immersive social group activities (Bernard Marr & Co, 2022).

Although social media has the advantage of facilitating connections with loved ones, it has also been criticized for promoting detrimental actions such as spreading fake news, conspiracy theories, harassment, and cyberbullying (Forbes Digital Assets, 2022).

A new and more immersive social media will make the process more engrossing and entertaining but it will also have the potential to magnify these threats. This can make Web3 version of social media a dangerous place. To venture into this realm, one must acknowledge and manage the risks that come with it, and be aware of the safety measures implemented by platform providers to mitigate these risks. For instance, Meta responded to reports of inappropriate conduct, such as "virtual groping," by introducing a "safe zone" function that permits users to promptly establish a protective boundary around themselves (Adgully, 2022).

References

101 Blockchain. (2022). *10 best metaverse platforms that you can try in 2022.* https://101blockchains.com/best-metaverse-platforms/. Accessed 21 Nov 2022.

Accenture. (2021). *25 cloud trends for 2021 and beyond.* https://www.accenture.com/nl-en/blogs/insights/cloud-trends. Accessed 22 Nov 2022.

Accenture. (2022a). *Government enters the metaverse.* https://www.accenture.com/content/dam/accenture/final/industry/public-service/document/Accenture-Federal-Technology-Vision-2022-Government-Enters-the-Metaverse New.pdf#zoom=40. Accessed 19 Dec 2022.

Accenture. (2022b). *Protecting and serving in the metaverse continuum.* Available at https://www.accenture.com/us-en/blogs/voices-public-service/public-safety-tech-vision. Accessed 20 Dec 2022.

Accenture. (2022c). *The next world after this: Aerospace and defence enters the metaverse.* https://www.accenture.com/_acnmedia/PDF-178/Accenture-Aerospace-Defense-Enters-Metaverse.pdf. Accessed 20 Dec 2022.

Accenture. (2022d). *Want to demystify the metaverse hype? Think of it as an internet evolution.* https://www.accenture.com/us-en/blogs/accenture-research/want-to-demystify-the-metaverse-hype-think-of-it-as-an-internet-evolution. Accessed 17 Oct 2022.

Accenture. (2022e). *Why the metaverse is a big gamechanger for defence.* https://www.accenture.com/us-en/blogs/voices-public-service/why-the-metaverse-is-a-big-gamechanger-for-defence. Accessed 20 Dec 2022.

Accenture. (2022f). *Meet me in the metaverse.* TechVision. https://www.accenture.com/_acnmedia/Thought-Leadership-Assets/PDF-5/Accenture-Meet-Me-in-the-Metaverse-Full-Report.pdf. Accessed 8 Oct 2022.

Accenture. (2022g). *Metaverse continuum set to redefine how the world operates.* https://www.accenture.com/us-en/blogs/intelligent-operations-blog/metaverse-continuum-set-to-redefine-how-the-world-operates. Accessed 21 Oct 2022.

Accenture. (2022h). *Meet me in the metaverse.* https://www.accenture.com/_acnmedia/Thought-Leadership-Assets/PDF-5/Accenture-Meet-Me-in-the-Metaverse-Full-Report.pdf. Accessed 21 Nov 2022.

Accenture. (2022i). *Why the metaverse (really) matters for travel.* https://www.accenture.com/us-en/blogs/compass-travel-blog/metaverse-travel. Accessed 21 Nov 2022.

Adgully. (2022). *The evolving face of Social Media—From socialising to the Metaverse.* https://www.adgully.com/the-evolving-face-of-social-media-from-socialising-to-the-metaverse-119631.html. Accessed 22 Nov 2022.

Analytics Insight. (2022a). *Welcome to the new world of art and culture with metaverse.* https://www.analyticsinsight.net/welcome-to-the-new-world-of-art-and-culture-with-metaverse/. Accessed 20 Dec 2022.

Analytics Insights. (2022b). *The metaverse—Bold plans for 2022 with Axie Infinity (AXS), and SeeSaw Protocol (SSW).* https://www.analyticsinsight. net/the-metaverse-bold-plans-for-2022-with-axie-infinity-axs-and-seesaw-pro tocol-ssw/. Accessed 20 Nov 2022.

Appinventiv. (2022). *How could metaverse be a game changer for the virtual gaming industry?* https://appinventiv.com/blog/metaverse-gaming/. Accessed 22 Nov 2022.

Bernard Marr & Co. (2022). *The future of social media in the metaverse.* https:// bernardmarr.com/the-future-of-social-media-in-the-metaverse/. Accessed 22 Nov 2022.

Blockchain Council. (2022a). *Decentraland metaverse: A complete guide.* https://www.blockchain-council.org/metaverse/decentraland-metaverse/. Accessed 22 Nov 2022.

Blockchain Council. (2022b). *Sandbox metaverse: An ultimate guide.* https:// www.blockchain-council.org/metaverse/sandbox-metaverse/. Accessed 21 Nov 2022.

Blockchain Council. (2022c). *Understanding the seven layers of the metaverse technology.* https://www.blockchain-council.org/metaverse/seven-lay ers-of-the-metaverse-technology/. Accessed 6 Nov 2022.

Blockchain Works. (2021). *Axie infinity developer scores $152M in series B funding, nearing $3B valuation.* https://blockworks.co/news/axie-infinity-developer-scores-152m-in-series-b-funding-nearing-3b-valuation. Accessed 21 Nov 2022.

Blockworks. (2022). *Metaverse platforms set the record straight about daily active users.* https://blockworks.co/news/metaverse-platforms-set-the-record-straight-about-daily-active-users. Accessed 22 Nov 2022.

Bloomberg. (2021). *Metaverse may be $800 billion market, next tech platform.* https://www.bloomberg.com/professional/blog/metaverse-may-be-800-bil lion-market-next-tech-platform/. Accessed 21 Nov 2022.

Bloomberg. (2022). *Apple shows AR/VR headset to board in sign of progress on key project.* https://www.bloomberg.com/news/articles/2022-05-19/apple-shows-headset-to-board-in-sign-it-s-reached-advanced-stage?leadSource=uve rify%20wall. Accessed 20 Nov 2022.

Business Insider. (2022). *AmEx reveals its metaverse ambitions in a trademark filing for tech to let people to use its payment cards in virtual worlds.* https:// www.businessinsider.in/cryptocurrency/news/amex-reveals-its-metaverse-ambitions-in-a-patent-filing-for-tech-to-let-people-to-use-its-payment-cards-in-virtual-worlds/articleshow/90236616.cms. Accessed 22 Nov 2022.

Business to Community. (2022). *Disney to adopt a metaverse platform to enhance the future of storytelling.* https://www.business2community.com/nft-news/ disney-to-adopt-metaverse-in-enhancing-the-future-of-storytelling-02548198. Accessed 22 Nov 2022.

Business Today. (2022). *India's first 'Metaversity' seems to be imploding from within; here's what's going on.* https://www.businesstoday.in/technology/news/story/indias-first-metaversity-seems-to-be-imploding-from-within-heres-whats-going-on-334891-2022-05-24. Accessed 22 Nov 2022.

Chengoden, R., Victor, N., Huynh-The, T., Yenduri, G., Jhaveri, R. H., Alazab, M., ..., Gadekallu, T. R. (2022). Metaverse for healthcare: A survey on potential applications, challenges and future directions. *arXiv preprint*, arXiv:2209.04160.

Coin Desk. (2021). *How Axie infinity creates work in the metaverse.* https://www.coindesk.com/markets/2021/07/17/how-axie-infinity-creates-work-in-the-metaverse/. Accessed 21 Nov 2022.

Coin Telegraph. (2021). *Microsoft metaverse vs. Facebook metaverse: What's the difference?* https://cointelegraph.com/metaverse-for-beginners/microsoft-metaverse-vs-facebook-metaverse-what-is-the-difference. Accessed 22 Nov 2022.

Coin Telegraph. (2022). *The feds are coming for the metaverse, from Axie Infinity to Bored Apes.* https://cointelegraph.com/news/the-feds-are-coming-for-the-metaverse-from-axie-infinity-to-bored-apes. Accessed 20 Nov 2022.

Decentraland. (2022). *Decentraland public launch.* https://decentraland.org/blog/announcements/decentraland-announces-publich-launch/. Accessed 22 Nov 2022.

Deloitte. (2021a). *The journey to government's digital transformation.* https://www2.deloitte.com/content/dam/insights/us/articles/digital-transformation-in-government/DUP_1081_Journey-to-govt-digital-future_MASTER.pdf. Accessed 21 Oct 2022.

Deloitte. (2021b). The metaverse overview: Vision, technology, and tactics. https://www2.deloitte.com/content/dam/Deloitte/cn/Documents/technology-media-telecommunications/deloitte-cn-tmt-metaverse-report-en-220304.pdf. Accessed 22 Nov 2022.

Digite. (2021). Remote work in the metaverse—How will it change the way we work? https://www.digite.com/blog/metaverse-remote-work/. Accessed 22 Nov 2022.

Duan, H., Li, J., Fan, S., Lin, Z., Wu, X., & Cai, W. (2021). Metaverse for social good: A university campus prototype. In *Proceedings of the 29th ACM International Conference on Multimedia* (pp. 153–161).

Economic Times. (2022a). *Now Shemaroo offers movie theatre on metaverse.* https://economictimes.indiatimes.com/tech/technology/now-shemaroo-offers-movie-theatre-on-metaverse/articleshow/94538342.cms?from=mdr. Accessed 6 Nov 2022.

Economic Times. (2022b). *Rise of metverse in global Pharma industry.* https://health.economictimes.indiatimes.com/news/pharma/rise-of-metaverse-in-global-pharma-industry/95250523. Accessed 22 Nov 2022.

End Points News. (2022). *The medical metaverse is already here, but what does that mean for pharma?* https://endpts.com/the-medical-metaverse-is-already-here-but-what-does-that-mean-for-pharma/. Accessed 21 Nov 2022.

Express VPN. (2022). *Survey reveals surveillance fears over the metaverse workplace.* https://www.expressvpn.com/blog/survey-reveals-surveillance-fears-over-the-metaverse-workplace/. Accessed 21 Nov 2022.

Facebook. (2021). Introducing horizon workrooms: Remote collaboration reimagined. https://about.fb.com/news/2021/08/introducing-horizon-workrooms-remote-collaboration-reimagined/. Accessed 22 Nov 2022.

Fintech Magazine. (2022). *JP Morgan is first leading bank to launch in the metaverse.* https://fintechmagazine.com/banking/jp-morgan-becomes-the-first-bank-to-launch-in-the-metaverse. Accessed 21 Nov 2022.

Forbes. (2022a). The challenges and opportunities with the metaverse. https://www.forbes.com/sites/forbestechcouncil/2022/05/17/the-challenges-and-opportunities-with-the-metaverse/?sh=38834232495f. Accessed 3 Jan 2022.

Forbes. (2022b). *The future of social media in the metaverse.* Enterprise Tech. https://www.forbes.com/sites/bernardmarr/2022/08/24/the-future-of-social-media-in-the-metaverse/?sh=3e4e24011023. Accessed 6 Oct 2022.

Forbes. (2022c). 6 top metaverse coins. https://www.forbes.com/advisor/investing/cryptocurrency/top-metaverse-coins/. Accessed 20 Nov 2022.

Forbes. (2022d). A short history of the metaverse. https://www.forbes.com/sites/bernardmarr/2022/03/21/a-short-history-of-the-metaverse/?sh=6e64a9ad5968. Accessed 6 Nov 2022.

Forbes. (2022e). *Disney: The metaverse, digital transformation, and the future of storytelling.* https://www.forbes.com/sites/bernardmarr/2022/10/07/disney-the-metaverse-digital-transformation-and-the-future-of-storytelling/?sh=19849efc13c0. Accessed 21 Nov 2021.

Forbes. (2022f). *Meta's VR vs. Apple's AR strategy—Who will ultimately win?* https://www.forbes.com/sites/timbajarin/2022/10/11/metas-vr-vs-apples-ar-strategy-who-will-ultimately-win/?sh=7ae676bb44ed. Accessed 21 Nov 2022.

Forbes Digital Assets. (2022). *The world of metaverse entertainment: Concerts, theme parks, and movies.* https://www.forbes.com/sites/bernardmarr/2022/07/27/the-world-of-metaverse-entertainment-concerts-theme-parks-and-movies/?sh=498e20176531. Accessed 21 Nov 2022.

Forbes India. (2022). *What will learning in the metaverse look like?* https://www.forbesindia.com/article/take-one-big-story-of-the-day/what-will-learning-in-the-metaverse-look-like/77285/1. Accessed 22 Nov 2022.

Gizmodo. (2022). *Employees are much more concerned about working in the 'metaverse' than their boss is.* https://gizmodo.com/metaverse-vr-meta-work-from-home-remote-work-1849342330. Accessed 21 Nov 2022.

Globe Newswire. (2022). *Cryptovoxels is rebranding to voxels on May 3, 2022.* https://www.globenewswire.com/news-release/2022/05/03/2434939/0/en/Cryptovoxels-Is-Rebranding-to-Voxels-on-May-3-2022.html. Accessed 21 Nov 2022.

HR TechX. (2022). *Future is here: How metaverse becomes the part of HR technology.* https://www.hrtechx.com/2022/02/15/future-is-here-how-metaverse-becomes-the-part-of-hr-technology/. Accessed 21 Nov 2022.

Inc. (2022). *According to Apple CEO Tim Cook, the next internet revolution is not the metaverse: It's this.* https://www.inc.com/nick-hobson/apple-ceo-tim-cook-next-internet-revolution-this-1-thing-metaverse.html. Accessed 21 Nov 2022.

Inc 42. (2022). *How metaverse is reinventing healthtech & its future.* https://inc42.com/resources/how-metaverse-is-reinventing-healthtech-its-future/. Accessed 22 Nov 2022.

Leeway Hertz (2022a). *Digital twin and metaverse.* https://www.leewayhertz.com/digital-twin-and-metaverse/. Accessed 8 Oct 2022.

Leeway Hertz. (2022b). *Metaverse: Uplifting the virtual gaming.* https://www.leewayhertz.com/gaming-in-metaverse/. Accessed 22 Nov 2022.

Live Mint. (2022). *India's leading online school: 21k school is delivering the future of education.* https://www.livemint.com/brand-stories/indias-leading-online-school-21k-school-is-delivering-the-future-of-education-11654524293651.html. Accessed 21 Nov 2022.

McKinsey. (2022). *Building a safer metaverse.* https://www.mckinsey.com/capabilities/growth-marketing-and-sales/our-insights/building-a-safer-metaverse. Accessed 21 Nov 2022.

Medium. (2021). *The metaverse value-chain.* https://medium.com/building-the-metaverse/the-metaverse-value-chain-afcf9e09e3a7. Accessed 6 Nov 2022.

Meta Cat. (2022). *Metaverse analytics.* https://www.metacat.world/en-US/analytics?typoe=cryptovoxels. Accessed 21 Nov 2022.

Meta Madrill. (2022). *Apple metaverse strategy; Apple's strategy for the digital universe.* https://metamandrill.com/apple-metaverse-strategy/. Accessed 21 Nov 2022.

Metahero. (2022). *Metahero.* https://metahero.io/uploads/Metahero_WP_v3_4.pdf. Accessed 21 Nov 2021.

Nasdaq. (2022). *Disney is diving into web 3.0 and the metaverse: Here's what that means.* https://www.nasdaq.com/articles/disney-is-diving-into-web-3.0-and-the-metaverse%3A-heres-what-that-means. Accessed 21 Nov 2022.

News 18. (2021). *What is metaverse and why Facebook/meta thinks it's the future of internet.* https://www.news18.com/news/tech/explained-what-is-metaverse-and-why-facebook-meta-thinks-its-the-future-of-internet-4416881.html. Accessed 21 Nov 2022.

Newzoo. (2020). *The world's 2.7 billion gamers will spend $159.3 billion on games in 2020; the market will surpass $200 billion by 2023.* https://newzoo.com/insights/articles/newzoo-games-market-numbers-revenues-and-audience-2020-2023. Accessed 21 Nov 2022.

PC Mag. (2022). *What is microsoft's metaverse strategy?* https://www.pcmag.com/news/what-is-microsofts-metaverse-strategy. Accessed 22 Nov 2022.

Phemex. (2022). *What is metahero: The gateway to the metaverse.* https://phemex.com/academy/what-is-metahero-hero-coin. Accessed 21 Nov 2022.

Revfine. (2022). *How the metaverse will change the travel industry.* https://www.revfine.com/metaverse-travel/. Accessed 22 Nov 2022.

Rospigliosi, P. A. (2022a). Adopting the metaverse for learning environments means more use of deep learning artificial intelligence: This presents challenges and problems. *Interactive Learning Environments, 30*(9), 1573–1576.

Rospigliosi, P. A. (2022b). Metaverse or simulacra? Roblox, Minecraft, Meta and the turn to virtual reality for education, socialisation and work. *Interactive Learning Environments, 30*(1), 1–3.

Tech Story. (2022). *How to access voxels metaverse (formerly cryptovoxels)?* https://techstory.in/how-to-access-voxels-metaverse-formerly-cryptovoxels/. Accessed 21 Nov 2022.

The Business Journals. (2022). *Here's how Accenture is manifesting the metaverse.* https://www.bizjournals.com/austin/news/2022/06/20/here-s-how-accenture-is-manifesting-the-metaverse.html. Accessed 22 Nov 2022.

The Conversation. (2022). *Tourism and the metaverse: Towards a widespread use of virtual travel?* https://theconversation.com/tourism-and-the-metaverse-towards-a-widespread-use-of-virtual-travel-188858. Accessed 21 Nov 2022.

The Verge. (2021). *Meta's sci-fi haptic glove prototype lets you feel VR objects using air pockets.* https://www.theverge.com/2021/11/16/22782860/meta-facebook-reality-labs-soft-robotics-haptic-glove-prototype. Accessed 22 Nov 2022.

Thomason, J. (2021). MetaHealth-how will the metaverse change health care? *Journal of Metaverse, 1*(1), 13–16.

Travel Dine. (2022). *Six ways metaverse will impact travel and hotels.* https://www.traveldine.com/six-ways-metaverse-will-impact-travel-and-hotels/. Accessed 20 Nov 2022.

Verdict. (2022). *Disney patent proves it is readying itself for the metaverse.* https://www.verdict.co.uk/disney-metaverse-patent/. Accessed 22 Nov 2022.

Wunderman Thompson. (2022). *Banking in the metaverse.* https://www.wundermanthompson.com/insight/banking-in-the-metaverse. Accessed 22 Nov 2022.

XR Today. (2021a). *Accenture orders record 60,000 Oculus headsets.* https://www.xrtoday.com/virtual-reality/accenture-orders-record-60000-oculus-headsets/. Accessed 21 Nov 2022.

XR Today. (2021b). *Gaming in the metaverse: The next frontier?* https://www.xrtoday.com/virtual-reality/gaming-in-the-metaverse-the-next-frontier/amp/. Accessed 21 Nov 2022.

Xu, M., Ng, W. C., Lim, W. Y. B., Kang, J., Xiong, Z., Niyato, D., ..., Miao, C. (2022). A full dive into realizing the edge-enabled metaverse: Visions, enabling technologies, and challenges. *IEEE Communications Surveys & Tutorials, 25*(1), 656–700.

Xu, X., Zou, G., Chen, L., & Zhou, T. (2022a). Metaverse space ecological scene design based on multimedia digital technology. *Mobile Information Systems, 2022.*

Zeb Pay. (2022). *Banking in metaverse—The future of banking.* https://zebpay.com/in/blog/banking-in-the-metaverse. Accessed 21 Nov 2022.

Zhang, M., Zhang, Z., Chang, Y., Aziz, E. S., Esche, S., & Chassapis, C. (2018). Recent developments in game-based virtual reality educational laboratories using the microsoft kinect. *International Journal of Emerging Technologies in Learning (iJET), 13*(1), 138–159.

Zuckerberg, M., & King, G. (2021). *Facebook launches "horizon workrooms." Here's how it works.*

CHAPTER 5

Metaverse for Public Sector

Abstract The government has not started to leverage this emerging technology of Metaverse and has been actively participating in incorporating digital transformation. This chapter discusses how government has been adopting digital maturity with the help of five levels. Further, the chapter mentions multiple barriers and challenges that the government has been facing in this course. Its application and implication are duly discussed along with the current public policy accentuated around the metaverse by the government.

Keywords Broadcasting model · Critical flow model · Data-centric · Mobilization and lobbying model · Comparative analysis model

5.1 Evolution and Digital Maturity of Government

As per the Deloitte survey 2021, governments around the world are at various stages of digital transformation, with only a small percentage considered to be mature in their journey. The majority of governments are still in the early or developing stages. Out of the surveyed organizations, only 30% assessed their digital capabilities as being ahead of their public sector peers, while nearly 70% stated that they lagged behind the

© The Author(s), under exclusive license to Springer Nature   91
Singapore Pte Ltd. 2023
R. Gupta and S. K. Pal, *Introduction to Metaverse*,
https://doi.org/10.1007/978-981-99-7397-2_5

private sector. The report also found that overall satisfaction with the current reaction of organizations to digital trends and their confidence in readiness to respond to digital trends were low.

Five levels of maturity model help in plotting their digital strategy and communicating it to key stakeholders and policymakers in the public sector (Gartner, 2017). All five levels are discussed in detail in the further sections.

5.1.1   Level 1: Initial (E-Government)

The current focus at this level is on transitioning services online to improve convenience and reduce costs, but the usage of data is limited and flawed. If organizations view a high proportion of online services or mobile access as the definition of a modern digital government, there is a need for further education and advocacy to demonstrate the true nature of a digital government and its benefits to both the government and the public (Lv et al., 2022). To make a case for progress, it is important to create case studies that illustrate how digital transformation can alleviate or eliminate high-priority pain points within the organization.

5.1.2   Level 2: Developing (Open)

According to Lee and Gu (2022), level 2 doesn't have to necessarily come after Level 1. E-government and open government programmes can exist together, each with their own objectives and management. Open government initiatives are typically designed to increase transparency, encourage citizen involvement, and promote the use of data. Examples of open data projects can be seen in smart city initiatives such as the Copenhagen Data Exchange.

5.1.3   Level 3: Defined (Data-Centric)

At this stage, the focus moves beyond just meeting the needs of citizens or users to proactively exploring new opportunities by strategically collecting and using data. According to Wang et al. (2022b), the performance indicators at this level are "how much data is open?" and "how many applications are developed using open data?" While it may be tempting to jump ahead without laying the proper groundwork, it is important to stay focused on designing and implementing strategies that are data-driven.

5.1.4   Level 4: Managed (Fully Digital)

In this stage, the government entity has fully embraced a data-centric strategy to improve its services and prefers open data principles as a means of innovation. This leads to the smooth exchange of data between organizations and improves services for citizens. However, there may be concerns regarding privacy, as citizens may feel uneasy about how their data is being collected and utilized. It is, therefore, crucial to adhere to existing norms and regulations regarding data usage and ensure that they are transparently communicated to the public. Qi and Chen (2022) suggest that this stage can encounter some privacy-related challenges, and hence, it is important to handle them carefully.

5.1.5   Level 5: Optimizing (Smart)

In this stage, the utilization of open data for digital innovation is deeply ingrained across the entire government, with support and direction from high-level policymakers. The innovation process is reliable and can be replicated consistently, even in the presence of unexpected events that necessitate swift action (Stanoevska-Slabeva, 2022).

5.2   CHALLENGES FOR GOVERNMENT

The current data protection law is regulated by specific regulations, including the Information Technology Rules of 2011 under the Information Technology Act 2000. To adhere to the guidelines in this provision, an organization needs to exhibit code compliance through information security policies and detailed security plans that incorporate operational, technical, and physical security measures (Theis & Wong, 2017). The legal and regulatory considerations consist of five main areas, viz., concerns regarding privacy and security, infringements of intellectual property rights, application of antitrust and competition laws, implementation of liability, and jurisdictional concerns.

The lack of specific regulations in India for overseeing metaverse raises concerns and poses risks to users. The absence of appropriate legal frameworks creates a void in the safe execution of metaverse (Mondaq, 2022). It is crucial for regulators to intervene at this stage while the technology is still in its infancy, as it will be more difficult to oversee once there is greater user dependence. Although innovative technologies such as AR,

VR, blockchain, and others have potential applications, their impact on established legal systems may cause scepticism.

It is evident that governing new technologies would be a challenging task and would require a revamping of existing legal systems. Additionally, insufficient funding poses a significant obstacle to digital transformation. Many public entities find it difficult to fund essential citizen services, let alone invest solely in digital initiatives, even though digital transformation is widely acknowledged as a way to achieve substantial cost savings.

According to Accenture (2022c), although 82% of organizations view digital technology as an opportunity, only 44% were able to increase their investments in such initiatives during the last fiscal year. Leaders who are attempting to drive change face conflicting priorities, particularly in striking a balance between transformation and maintaining current operations. Successful public bodies are those with well-defined business cases and coherent strategies. Mature organizations, in addition to many competing priorities and insufficient funding, often cite security as a significant barrier. In contrast, early-stage agencies' primary barriers include a lack of an overall strategy and limited comprehension of digital trends.

The dimension of workforce skills is a challenging aspect of digital transformation, with culture being the second most challenging. While there is recognition of the need to have a digital-savvy workforce, there is an understanding that changing culture is a difficult task. Addressing the skills gap by hiring people with the necessary skills or providing training for current employees can help in dealing with this challenge.

A survey conducted among public sector leaders revealed that most government organizations lack the strategy to achieve digital transformation. On one hand, strategy is the bedrock of the transformation process. However, on the other hand, the leaders do not even realize its significance or importance. More than half of the public organizations that are currently functioning in the Indian public sector lack digital strategy. As the metaverse is anticipated to be accessed by individuals from different nations, backgrounds, and geographical locations, questions arise regarding which countries' laws will govern the digital realm and metaverse environment. With the absence of physical borders in the virtual space, determining jurisdiction is even more uncertain and has become a major concern for various government entities (Mondaq, 2022).

The government's stance on metaverse and digital maturity is uncertain, and no definitive answer has been given. The best course of action may be to merge and create legislation that adapts to the shifting landscape, incorporating various provisions to address new challenges and offer solutions to the world.

5.3 METAVERSE AS A SOLUTION: ROLE AND DIFFERENT MODELS

The government can take significant advantage of plenty of benefits offered by the metaverse. Metaverse has the capability to make services and applications available to citizens through avatars whose availability is 24×7. Especially during the pandemic, it restricted the mobility of people which possessed a challenge to citizens to use applications and services. The metaverse could serves as a platform to offer applications and services to the citizens using a technology replicating the 3D world in which people live in.

Government across the world such as Sweden, Maldives, Serbia, and Estonia have already established their virtual embassies. Many governments have also adopted virtual simulations for interactive learning, voting, and conferences. The US Federal, state, and local governments have particularly notable and elaborate examples of this. One of the government local examples is MuniGov. This refers to a partnership between state, federal, municipal, and international governments that are investigating the application and principles of Web 2.0 to enhance communication and citizen services through technology. As an example, a virtual polling station was established in Alameda County, California, USA to educate people on voting procedures using technology. India is also promoting metaverse and some bits and pieces of its application can also be noticed in the Indian context as well. The Telangana government has unveiled its Space Tech Framework with ISRO on PartyNite, a metaverse platform created by Gamitronics.

With the private sector aggressively promoting the adoption of the metaverse, the government should establish appropriate policies and utilize the metaverse to enhance public services. In the past, the government successfully utilized Web 2.0 to provide online services, and now a similar opportunity exists to leverage the advances in Information and Communication Technology.

5.3.1   Broadcasting Model

The broadcasting approach involves using ICT to spread useful information related to governance. The government has the option of creating a meta-division within existing broadcasting agencies like All India Radio, Lok Sabha TV, Doordarshan, Rajya Sabha TV or establishing its own meta-gov.

In the coming years, the "Meta Mann ki Baat" could allow people to interact with a digital version of the Prime Minister, while the "Pariksha pe Charcha" metaverse could provide students, teachers, and parents with a similar opportunity. Unlike the broadcasting model, where people are limited to a one-way channel and unable to interact or ask questions from authorities, these metaverse platforms would allow for more engagement and interaction.

5.3.2   Critical Flow Model

According to Bala and Verma (2018), the critical-flow model of e-governance allows the release of essential information to the targeted audience through ICT. The establishment of a meta-help desk or meta-division within a government agency could facilitate the provision of essential data. During a meta-meet, the government could inform organizations like the Indian Medical Association (IMA) about insurance coverage for doctors, and disseminate updates about the pension scheme to pensioners in the meta-room.

5.3.3   Interactive Services Model

This approach facilitates bidirectional exchange of information between participants and is primarily employed to offer public services online via a Government to Consumer to Government (G2C2G) Model.

A Regional Transport Office (RTO) can easily inspect the vehicle and submit documents while sitting at home, plugging into VR headsets in the meta-RTO in the future (Niti Ayog, 2022). The public official's avatars in the metaverse will initiate civil services and convenient consultations that are available only through Municipal corporations. The ecosystem interoperability is expected to be such that the same user will be able to easily move through the municipal corporation, passport office, and RTO without multiple log-ins to individual platforms. It can become a unified and continuous experience for the citizens.

5.3.4   Mobilization and Lobbying Model

Algazo et al. (2021) explain that the e-advocacy or mobilization and lobbying model aims to establish online forums and collect public opinions for a specific policy. By doing so, this model facilitates the formation of various virtual communities and enables the accumulation of ideas, resources, and expertise through online networking. This approach allows for the utilization of human resources that extend beyond bureaucratic and geographical constraints.

In the future, officials from the Ministry of Commerce may engage in virtual meetings with the Federation of Indian Chamber of Commerce and Industry (FICCI) to formulate business policies. Likewise, representatives from the Ministry of Law could hold virtual meetings with the Bar Council of India (BCI), and the Ministry of Education could have virtual meetings with the All India Federation of University and College Teachers (AIFUCT) for efficient policy creation (Algazo et al., 2021).

5.3.5   Comparative Analysis Model

The comparative analysis model is of great importance for nations in the process of development, as it has the potential to empower individuals. This model involves creating benchmark indicators such as the School Education Quality Index and the Innovation Index. These benchmarks are then used to compare regional parameters at the district, state, and national levels (Grigalashvili, 2022). In a virtual environment, the Prime Minister and Chief Ministers of all states can work together in what is referred to as a "meta-meet." During this meeting, each Chief Minister can discuss their current governance practices. By comparing data over a period of time, the Chief Ministers can gain an understanding of the past and present state of specific, pre-determined benchmarks such as the Infant Mortality Rate (IMR) and Maternal Mortality Rate (MMR).

5.4   POTENTIAL USE-CASES
OF METAVERSE FOR GOVERNMENT

5.4.1   Aerospace and Defence

The global AR and VR investment in Aerospace and Defence investment is expected to reach USD 5.9 billion by 2025 (Devden, 2022). The increasing significance of Augmented Reality (AR) and Virtual Reality (VR) in the aerospace and defence industry is demonstrated by the number of metaverse-related patents that have been published. Thales SA holds the largest number of such patents globally in the Aerospace and Defence sector, with 156 patents published between 2002 and 2022. LG Corp has the second-highest number of metaverse patents, with 138. The third largest number of patents was published by The Boeing Corp, with 86, followed by Airbus SE with 69 patents, with 37.7% of those patents being contributed by its subsidiary Airbus Operations SAS. Finally, BAE Systems Plc has the fifth largest number of patents, with 48 (Global Data, 2022).

Research revealed that 70% of public service executives believe that the metaverse would bring greater benefits to the public sector. 50% of them believe that it will have a breakthrough impact in the next four years. The four areas where the defence industry can use metaverse to drive new value are:

- Retaining and attracting talent
- Training and simulation
- Command and control
- Procurement and supply chain.

Defence organizations functioning worldwide are facing real concerns about talent pipelines. The metaverse can help in providing an exciting pathway for the hyper-personalization of recruitment to drive engagement (Madou, 2022). It can provide potential recruits with a taste of what military life can be and provide a real-like compelling experience that could attract potential recruits. In parallel to this, AI can broaden the talent pool where people from different communities and demographics can be recruited. This type of recruitment can also be called a hidden workforce. The intelligent search could act as a key to help self-defence in competing with the top-talents.

According to Pozniak (2022), the other aspect can be staff retention and enhancement in the work environment. For instance, armed-forces personnel often work across multiple systems simultaneously. It means using one system for keeping a check over what is going on in the field and the other system for the diagnostic of equipment failure, tracking repairs, and ordering parts. Metaverse can help in combining all these systems in a single, unified, and compelling experience. It in turn can help in driving employee engagement and retention.

The defence industry has been using virtual reality for decades (Accenture, 2022c). Usually, it is used in small environments for specific purposes, such as raining someone to operate a plane, submarine, or tank. But, these simulations have outpaced developments in various other scenarios also neither they are immersive nor integrated. Metaverse can help in taking these simulations to an altogether different level. It can empower organizations to create a highly realistic virtual world that can prepare the personnel better for challenging situations. Creating a common synthetic environment for group training or individuals and transporting trainees in simulated realistic situations can offer powerful ways to prepare the recruited for the upcoming challenges.

The metaverse has the potential to significantly enhance national security through improved interoperability and more effective collaboration with allies (Accenture, 2022c). By utilizing the metaverse, commanders can make informed decisions in real-time and minimize operational risks. For example, a defence organization that is scheduled to receive a new type of aircraft in 18 months can use the metaverse to assess the potential impact on their overall defence capabilities. According to Dionisio and Gilbert (2013), if the organization suspects an adversary has obtained a new weapon, the metaverse can be used to evaluate the possible effects and take necessary measures to mitigate them. Additionally, the metaverse can be used to quickly simulate various scenarios during the planning of an operation, allowing the organization to predict outcomes and determine the best course of action.

Additionally, defence organizations can harness the capabilities of the metaverse in high-risk situations to effectively integrate intelligence from various sources. The goal is to provide each commanding officer with tailored and up-to-date information at their disposal whenever they need to make a decision.

The metaverse also presents significant opportunities for procurement and supply chain operations in the defence sector. It can be utilized to enhance the security of supply by conducting comprehensive scenario planning. Further, the metaverse can also help organizations in trialling and selecting equipment. Besides this, it can also assist in optimising equipment allocation (Xu et al., 2022b). Technologies such as Web 3.0 and blockchain can aid organizations in accessing data, verifying its origin, and unlocking its potential value.

5.4.2   Art and Culture

Metaverse is emerging as a new medium that is integrating art and technology for a better purpose. Popular, as well as rising artists, can use this next-generation digital platform in creating and producing fascinating art with the collaboration of VR headsets. Metaverse platform is known for its ability to provide ample space and opportunities for bringing out inner talent and skills and earning profits with seamless transactions. According to Andreula and Petruzzelli (2022), the platform also allows artists to showcase their art and cultures to global audiences through virtual galleries that offer a seamless experience to them.

Cryptovoxels is a well-known Metaverse art among the communities of art galleries and museums such as the San Francisco Museum of Modern Art and the FC Francisco Carolinum Linz for users to store art galleries (Analytics Insight, 2022b). Apart from this, Cryptoweiner is also gaining popularity for generating art projects backed by blockchain technology for building and uploading art and cultural events on the metaverse. Besides this, Adobe is focused on new and emerging 3D projects for providing the future design tool for the metaverse.

The best way to preserve traditional culture is by giving intensive knowledge to people through books, museums, libraries, and photographs.

According to Bowen and Giannini (2022), about 2% of people, on an average visit a museum once a month and 15% visit them once annually. Roaming and appreciating art pieces in galleries is a surreal experience, but metaverse can help in making it more personalized. 3D and 360-degree views through VR headsets will allow people to have more life-like glimpses of the famous sites of the world. Metaverse will help in making these tours more elaborate. It can also help in increasing footfall by entertaining people through metaverse access apart from the physical museums and art galleries (Mohanty & Swain, 2022). For instance, the Indian Museum in Kolkata can have visitors from all around the world apart from those who are visiting India for tourism purposes. Hence, it will become easier to convey the Indian culture without any physical presence of the visitors.

With the metaverse, Art will become more fluid and interactive with the possibilities that virtual reality presents. Considering the growing inclination towards technology, cultural institutions must pay attention to growing trends of digital art and online digital sales (Jung, 2022).

For example, non-fungible tokens (NFTs), which are digital art forms, can fetch high prices when sold. On December 2, 2021, Pak's artwork "The Merge" was sold for $91.8 million. This is significant, as the estimated value of the widely renowned painting "The Starry Night" is approximately $100 million (Automatic Sync, 2022). Artists and members of the art community who have already adopted digital forms of art and utilized technology-powered platforms to expand their online reach will be best prepared for success in the forthcoming metaverse.

5.4.3   Trade and Economy

The trading industry stands to benefit significantly from emerging technological advancements, as complex products and rich datasets provide an ideal opportunity to leverage technologies such as Augmented Reality (AR) to support daily business operations. In particular, AR and Virtual Reality (VR) can create an unconstrained environment for visualizing data, allowing for easy recognition of changes and patterns, collaboration with clients, and interaction with counterparties in dynamic and real-time systems. This can lead to a reduction in overall business infrastructure costs, as multiple monitors, prime real estate, systematic cooling, and advanced wiring become unnecessary.

As market data providers continue to evolve their products, it will become increasingly important for them to develop Application Programming Interfaces (APIs) that are tailored for consumption via different AR and VR solutions. The Metaverse has the potential to revolutionize forex trading by eliminating borders, strict regulations, and trade barriers, allowing individuals from any country to trade without impediments. Will metaverse change the global trade? Short answer? Yes, a lot. Under the metaverse, global trade will be automated and permanently online. People will be able to access a global unified market using a single avatar entity, regardless of the entry point into the metaverse. The most obvious changes will be the consolidation of a global unified market, the auto-mated supply chain, and the empowerment of global culture (Tong, 2022). Global cultural means a product is appealing to all geographical regions at the same time. For instance, a product is compatible with people from China and Germany both, which apparently have an entirely different culture.

In short, the metaverse will bring robotization and automatization to the supply chain and delivery process. It will also fasten speed up the delivery process without any human limitations. Metaverse will ensure the integration of all the platforms and media into a single avatar developing global standards of shipping.

It is predicted that by 2023, the automation of warehouses will have grown by over 38% and by 2025, the majority of the world's warehouses will be largely, or totally automated and drone-operated (Trade Finance Global, 2022). The metaverse will transform global trade in to more efficient, automated process that operates 24 hours and 7 days a week. The elimination or decrease in the delays of accessibility of information will level up the playing field here some businesses might lose their competitive advantage from years of direct engagement with customers and suppliers. However, at the same time, it will make the process of finding new suppliers and customers easier. The procedure has already begun at the warehouse level.

Currently, only 20% of warehouses globally are fully automated, but this percentage is rapidly growing (Faraboschi et al., 2022). In the future, entire trading marketplaces and stock exchanges may be built in the meta-verse, providing investors with a much larger pool of opportunities than could ever be possible in the physical world, due to geographical and political limitations.

5.4.4   Employment

Metaverse has the capability to radically change how we work. Employees in the metaverse can be represented as avatars. However, it is impor-tant to note whether the employers will be able to see the avatars of the employees and reflect their age, physical appearance, ethnicity, gender, or disability or not (Carter, 2022). Also, it will be possible for employers to hold avatars responsible for their actions and conduct.

Employers have the duty of care for the employees for ensuring a safe working environment. It becomes difficult for employers to meet this requirement of duty of care in the virtual working space. In future, it can become difficult for the employer to evaluate whether the employer will be vicariously liable for the acts in the other virtual world or not (Bennett, 2022). In such cases, the employment rules demand reasonable provisions that can make metaverse a safe working place for both employers and employees.

There are already varied reports of virtual sexual harassment and inappropriate conduct in the metaverse (Cnbc, 2022). Due to these concerns, some employers are taking steps to distance measures and develop boundaries. It is also possible that employees may be subjected to less favourable treatment because of a deemed or assumed protected characteristic. For instance, if the employees have the flexibility to choose the appearance of their avatars, there can be a further risk of discrimination, cultural appropriation, and stereotyping.

Employment practices will need an upgradation on mindful practices that caters to the wider workplace culture. Whenever there is a significant change in the way work is done, there is a risk of creating a two-tier workforce. This could happen in the context of the metaverse, where some workers may be non-users or slow adopters, while others may be willing to embrace it and work effectively in a virtual world. This could potentially create a divide between those who are able to keep up with technological advancements and those who are not, leading to inequalities in the workforce. It is important for organizations to provide adequate training and support to all employees to ensure a smooth transition to the metaverse, and to avoid creating a disparity that could negatively impact productivity and employee morale (Upadhyay & Khandelwal, 2022). Employers also need to consider how working in the metaverse can affect the status of diversity and inclusion if certain demographics in the workforce are not comfortable with working in the virtual environment and feel alienated by the new ways of working.

Metaverse can help in bridging the gap between attracting better talent and placing them at the right job description (People Matters, 2022). Metaverse will make virtual recruitment fairs and other online job-related events common where employees can learn more about employers from a unified place. This will make the process more interactive, informative, human, and fun allowing the companies to lift up the experience of the candidates beyond the ordinary. Using metaverse, companies can create a fully immersive journey for the candidates in a virtual environment where their people's culture, values, mission, and products come to life. Metaverse can also allow companies to offer behind-the-scenes tours of the workplace to prospective employees (Harvard Business Review, 2022b). It will enable them to have a closer look at what their life is going to be once they join the organization.

The process of applying for a job has gone digital but still has flaws that negatively impact the candidate's experience. Currently, the job application process is lengthy, cumbersome, and restrictive with numerous screens and questions. The metaverse has the potential to streamline the process for both employers and candidates, making it faster and more user-friendly (Choi, 2022). Employers can collect information from applicants in a way that enhances their experience and results in more accurate and reliable assessments.

5.4.5   Public Health and Safety

The aim of the Ayushman Bharat Digital Mission (ABDM) by Indian Government is to maintain the continuity of the company's integrated digital health infrastructure, providing a national digital health system that is cost-effective, comprehensive, secure, accessible, and promotes universal health coverage (Jena et al., 2022). It will offer access to a wide range of information, data, and infrastructure services by using interoperable, open, standards-based digital systems and make sure that the personal health information is private, safe, and confidential. Metaverse focuses on unified presence and interoperability which can prove to be a boon for public health.

It is crucial for India to play a role in global health governance and fulfil its commitments under the United Nations 2030 Agenda for Sustainable Development. Metaverse can be a great promise to public health by infusing technology like AR, VR, Internet of Medical Devices (IoMD), AI intelligent cloud, Web 3.0, quantum computing, and edge with robots to give an altogether new direction to healthcare (Chauhan et al., 2022). It is expected that the adoption of the metaverse will help in making healthcare more accessible to people. The metaverse can facilitate collaboration between specialists in large hospitals and practitioners in smaller facilities, improving the accuracy and effectiveness of diagnosis and treatment through a system of graded evaluations.

According to a survey conducted by Accenture in 2022, 98% of participants from the public sector agreed that technology advancements are increasingly more dependable in guiding their organization's long-term strategy, compared to economic, political, or social trends. If there is one thing that every public safety agency should be thinking about, it is what advancing technology means for its organization and how it can turn these advances into an advantage. Public safety agencies need to view this new reality through a new lens. One focuses on the opportunities it offers to better protect and serve citizens. The other looks at the potential risks and threats that new technologies can present when used by bad actors. Across both lenses, legitimacy and trust are key (Danylec et al., 2022). When the physical and digital lines are blurred, attributes like fairness, privacy, transparency, freedom, and human intervention from bias are more important than ever.

70% of public service executives believe the metaverse will have a positive impact on their organizations, with 50% already saying it will be breakthrough or transformational. Essentially, the huge potential for VR is used by public sector agencies today (Damar, 2022). These agencies can provide immersive training from physical to logistical from entering a blazing building to using firearms. Further, it can build greater levels of empathy and understanding of unconscious bias.

A related opportunity is around conversational AI through virtual agents and chatbots. These are the tools that people are getting increasingly comfortable with. While they need to be used with care, chatbots can relieve the pressure on call handlers in non-emergency call centres. This type of technology has broader applications with conversational AI opening up further possibilities from enabling people on probation to get support 24/7, to help citizens access information and guidance (Danylec et al., 2022).

The majority (67%) of consumers anticipate that companies will leverage technology to address significant and intricate societal issues, as it would be advantageous for both themselves and their communities (Accenture, 2022c). However, this will require public safety organizations to collaborate and create new alliances and partnerships to access both technologies and expertise. 99% of public service executives are concerned about deep fakes and disinformation attacks. Just about every crime now has a digital component, making it vital that public safety agencies ensure their people have the right digital skills and capabilities to respond to ever more sophisticated cybercrime.

5.4.6   Public Entertainment

The world of metaverse and entertainment or "metavertainment" has become one of the most discussed topics in the metaverse world. Metaverse in the entertainment market is expected to reach $221.7 Billion, Globally by 2031 at a CAGR rate of 32.3% from 2022 to 2031 (PS Newswire, 2022).

Kim and Yoo (2021) suggest that a rise in spending on events, virtual concerts, and other tech-driven innovations aimed at enhancing audience engagement and supporting franchise expansion is meeting consumer demand and driving growth in the entertainment industry. The ability to experience a completely alternate reality, detached from the real world, in a unique environment, is highly appealing. Theme and amusement parks with expansive areas can attract visitors from all over the world, at no extra cost, offering a cost-effective and time-efficient option compared to having to travel long distances to these locations.

Entertainment also includes sports. This digital ecosystem has also led to a rise in the popularity of sports betting as a profitable application. Virtual reality often enables one to observe the game better to taking smart decisions even more efficiently. It is set to offer its own crypto wallet for more competitive entertainment and enticing people. Popular artists such as BTS, Imagine Dragons, and many more have already started virtual reality concerts for their loyal supporters (Niu & Feng, 2022). Thus, Metaverse is also set to provide lucrative and large venues to hold concerts and perform for a long time, similar to multiple real-life music concerts.

5.4.7   Knowledge Management

Web 3.0 services can connect users and computers for intensive knowledge generation and problem-solving activities. As a result of its vast processing capacity, Web 3.0 can deliver high-value services to enterprises and consumers due to its assertiveness and high customization (Upadhyay & Khandelwal, 2022).

According to Anderson and Rainie (2022), the transfer of tacit knowledge, which has always been a major challenge in knowledge management, can be addressed with the help of the metaverse. When an expert is asked to perform a complex task, they may not be able to fully articulate the steps involved or some of the unconscious processes. However, by having the expert perform the task in the metaverse, a wealth of data can be collected about contextual information, technical gestures, human behaviour, etc. This can be done by equipping the expert with VR gear, video cameras, or VR gloves, allowing them to describe their actions as they perform the task, which will be recorded and translated into text in real-time (Upadhyay & Khandelwal, 2022). The equipped experts can make the instructions clear and help in better capturing, better understanding, better transferring and modelling the experiential knowledge.

According to Wang et al. (2022d), the metaverse holds the potential to greatly enhance knowledge dissemination and management. It offers an opportunity to provide education and training to teams in various industries, regardless of geographical location, time, or resources. The metaverse is poised to revolutionize the way education and training is conducted, just as the internet transformed the way information was shared. Through the metaverse, a range of educational and training experiences can be made accessible, from teaching basic science concepts to school children, to training professionals in complex procedures, such as orthopedic surgery or welding. This ability to share experiences across distances will bring a new level of accessibility to education and training (CII Knowledge Summit, 2022).

5.5 IMPLICATIONS OF METAVERSE ADOPTION FOR GOVERNMENT

Rospigliosi (2022c) highlights that as the metaverse continues to evolve, it presents both new opportunities and challenges that must be taken seriously by the government. On the one hand, the metaverse can be used to create jobs, provide healthcare, and deliver education in entirely new ways. On the other hand, the metaverse will also present challenges in areas such as intellectual property, taxation, identity verification, disinformation, and regulatory compliance. The government must prepare to address these challenges in order to ensure that the metaverse can be utilized to its full potential. Some of the specific questions that need to be addressed include how to deal with NFT sales of national monuments, how to protect intellectual property rights, how to tax virtual companies, how to connect digital personas to natural persons, and how to ensure regulatory compliance and tax reporting.

According to Hassani (2022), safeguarding the metaverse can be a challenge since users can come from anywhere globally. Governments may consider forming a strategic futures group to assess and monitor the dangers and possibilities that evolving immersive technologies bring about in the short, mid, and long-term future. For instance, Singapore has created a Center for Strategic Futures that operates under the office of the Prime Minister.

In addition, as noted by Bibri (2022), due to the intricate regulatory issues posed by the metaverse, the government ought to consider creating a multi-disciplinary group for anticipatory regulation. This group could conduct research on the necessary regulations and work with the private sector to examine regulatory methods for the metaverse.

5.6   Public Policy for Metaverse

Metaverse has the capability to enhance businesses as well as public services and expand social and economic opportunities for individuals as well as organizations. Data Stewardship has become one of the major concerns in this aspect. The amount of information gathered through the process is huge and therefore it generates the requirement for strong data protection and privacy measures (Dick, 2021). To safeguard the privacy of users, it is important for them to have a clear understanding of how their biometric data, like eye and motion tracking, is utilized by companies. The companies collecting this data must have proper measures in place to keep it safe from unauthorized access by third parties.

As stated by Bibri et al. (2022), the use of these technologies requires particular attention to privacy issues. Hence, the need for a comprehensive national privacy law that sets appropriate boundaries and provides a solid foundation for future tech developments has been emphasized by various researchers. The privacy law should be technology-neutral to ensure its continued relevance in the future.

It is crucial to consider the potential risks and harm related to the use of metaverse technology and devices, beyond just privacy issues. These technologies can have both physical and emotional consequences, such as changing the user's perception of reality and increasing the likelihood of manipulation. For example, an augmented reality application that hides the view of oncoming traffic could result in physical harm, while it may also open up new avenues for harassment or defamation. It is important to address these safety and security concerns to ensure the responsible use of metaverse technology. Hence, there are valid concerns associated with mental and physical concerns due to the overlapping of virtual and physical reality.

Child safety is a major cause of concern where metaverse technologies will be more present in classrooms, homes and other aspects of life for all ages (Kanematsu et al., 2014). It generates the importance of establishing guardrails that protect the emotional and physical well-being of children across these experiences. Jeremy Bailenson from Stanford VHIL has expressed concern regarding the impact of immersive technologies on child safety, which he considers to be a major area of unknown consequences (Information Technology & Innovation Foundation, 2021).

In addition to public security, researchers argued that a significant impact on national security can also be noticed. Thomason (2022) highlights the potential dangers of not having proper security measures in place for immersive technologies. Adversarial actors could manipulate the reality-altering capabilities of these technologies for malicious purposes. The dangers are also exemplified by the rise of deep fakes which can create false recorded images or videos. Additionally, AR and VR can present fake realities in real-time, such as altering digital overlays to misinform military personnel or officials during a crisis.

Metaverse technologies are proliferating in every aspect of life from communication and entertainment to education and workforce development. According to various studies, these emerging technologies have the capability to bring about major changes, but they also present unique challenges related to safety, privacy, fairness, and equity that policymakers must address in comparison to existing technologies (Kshetri, 2022).

REFERENCES

Accenture. (2022a). *Government enters the metaverse*. https://www.accenture.com/content/dam/accenture/final/industry/public-service/document/Accenture-Federal-Technology-Vision-2022-Government-Enters-the-Metaverse New.pdf#zoom=40. Accessed 19 Dec 2022.

Accenture. (2022b). *Protecting and serving in the metaverse continuum*. https://www.accenture.com/us-en/blogs/voices-public-service/public-safety-tech-vision. Accessed 20 Dec 2022.

Accenture. (2022c). *The next world after this: Aerospace and defence enters the metaverse*. https://www.accenture.com/_acnmedia/PDF-178/Accenture-Aerospace-Defense-Enters-Metaverse.pdf. Accessed 20 Dec 2022.

Accenture. (2022d). *Want to demystify the metaverse hype? Think of it as an internet evolution.* https://www.accenture.com/us-en/blogs/accenture-res earch/want-to-demystify-the-metaverse-hype-think-of-it-as-an-internet-evo lution. Accessed 17 Oct 2022.

Accenture. (2022e). *Why the metaverse is a big gamechanger for defence.* https://www.accenture.com/us-en/blogs/voices-public-service/why-the-metaverse-is-a-big-gamechanger-for-defence. Accessed 20 Dec 2022.

Accenture. (2022f). *Meet me in the metaverse.* TechVision. https://www.accent ure.com/_acnmedia/Thought-Leadership-Assets/PDF-5/Accenture-Meet-Me-in-the-Metaverse-Full-Report.pdf. Accessed 8 Oct 2022.

Accenture. (2022g). *Metaverse continuum set to redefine how the world operates.* https://www.accenture.com/us-en/blogs/intelligent-operations-blog/metaverse-continuum-set-to-redefine-how-the-world-operates. Accessed 21 Oct 2022.

Accenture. (2022h). *Meet me in the metaverse.* https://www.accenture.com/_acnmedia/Thought-Leadership-Assets/PDF-5/Accenture-Meet-Me-in-the-Metaverse-Full-Report.pdf. Accessed 21 Nov 2022.

Accenture. (2022i). *Why the metaverse (really) matters for travel.* https://www.accenture.com/us-en/blogs/compass-travel-blog/metaverse-travel. Accessed 21 Nov 2022.

Algazo, F. A., Ibrahim, S., & Yusoff, W. S. (2021). Digital governance emergence and importance. *Management, 6*(24), 18–26.

Analytics Insight. (2022a). *Welcome to the new world of art and culture with metaverse.* https://www.analyticsinsight.net/welcome-to-the-new-world-of-art-and-culture-with-metaverse/. Accessed 20 Dec 2022.

Analytics Insights. (2022b). *The metaverse—Bold plans for 2022 with Axie Infinity (AXS), and SeeSaw Protocol (SSW).* https://www.analyticsinsight. net/the-metaverse-bold-plans-for-2022-with-axie-infinity-axs-and-seesaw-pro tocol-ssw/. Accessed 20 Nov 2022.

Anderson, J., & Rainie, L. (2022). *The metaverse in 2040.* Pew Research Centre.

Andreula, N., & Petruzzelli, S. (2022). Meta-soft power: Flipping the scales between art & culture. *RAISINA FILES,* 144.

Automatic Sync. (2022). *What the metaverse means for cultural institutions & their digital presence.* https://www.automaticsync.com/metaverse-means-cul tural-institutions-digital-presence/. Accessed 20 Dec 2022.

Bala, M., & Verma, D. (2018). Governance to good governance through e-Governance: A critical review of concept, model, initiatives & challenges in India. *International Journal of Management, IT and Engineering, 8*(10), 244–269.

Bennett, D. (2022). Remote workforce, virtual team tasks, and employee engage-ment tools in a real-time interoperable decentralized metaverse. *Psychosociolog-ical Issues in Human Resource Management, 10*(1), 78–91.

Bibri, S. E. (2022). The social shaping of the metaverse as an alternative to the imaginaries of data-driven smart cities: A study in science, technology, and society. *Smart Cities, 5*(3), 832–874.

Bibri, S. E., Allam, Z., & Krogstie, J. (2022). The metaverse as a virtual form of data-driven smart urbanism: Platformization and its underlying processes, institutional dimensions, and disruptive impacts. *Computational Urban Science, 2*(1), 1–22.

Bowen, J. P., & Giannini, T. (2022). *Digital experience in art and identity: The metaverse calls.*

Carter, D. (2022). Immersive employee experiences in the metaverse: Virtual work environments, augmented analytics tools, and sensory and tracking technologies. *Psychosociological Issues in Human Resource Management, 10*(1), 35–49.

Chauhan, V., Dumka, N., Hannah, E., Ahmed, T., & Kotwal, A. (2022). Recent initiatives for transforming healthcare in India: A political economy of health framework analysis. *Journal of Global Health Economics and Policy, 2*, e2022002.

Choi, H. Y. (2022). Working in the metaverse: Does telework in a metaverse office have the potential to reduce population pressure in megacities? Evidence from young adults in Seoul, South Korea. *Sustainability, 14*(6), 3629.

CII Knowledge Summit. (2022). *Leveraging the metaverse in knowledge management.* https://ciiknowledgesummit.in/wp-content/uploads/2022/04/Met averse-for-Knowledge-Management-Paper.pdf. Accessed 20 Dec 2022.

CNBC. (2022). *Employers see promise in a metaverse workplace: Employees are a little more sceptical.* https://www.cnbc.com/2022/08/09/employers-see-promise-in-metaverse-workplace-employees-are-skeptical.html. Accessed 20 Dec 2022.

Damar, M. (2022). What the literature on medicine, nursing, public health, midwifery, and dentistry reveals: An overview of the rapidly approaching metaverse. *Journal of Metaverse, 2*(2), 62–70.

Danylec, A., Shahabadkar, K., Dia, H., & Kulkarni, A. (2022). Cognitive implementation of metaverse embedded learning and training framework for drivers in rolling stock. *Machines, 10*(10), 926.

Devden. (2022). *Enhance warfare and training.* https://www.devdensolutions.com/aerospace-defence/. Accessed 20 Dec 2022.

Dick, E. (2021). *Public policy for the metaverse: Key takeaways from the 2021 AR/VR policy conference.* Information Technology and Innovation Foundation.

Dionisio, J. D. N., & Gilbert, R. (2013). 3D virtual worlds and the metaverse: Current status and future possibilities. *ACM Computing Surveys (CSUR), 45*(3), 1–38.

Faraboschi, P., Frachtenberg, E., Laplante, P., Milojicic, D., & Saracco, R. (2022). Virtual worlds (metaverse): From skepticism, to fear, to immersive opportunities. *Computer, 55*(10), 100–106.

Gartner. (2017). *5 levels of digital government maturity.* https://www.gartner. com/smarterwithgartner/5-levels-of-digital-government-maturity. Accessed 19 Dec 2022.

Global Data. (2022). *Global: Top metaverse patents holders in the aerospace and defence sector.* https://www.globaldata.com/data-insights/aerospace-and-def ence/global-top-metaverse-patents-holders-in-the-aerospace-and-defence-sec tor-2132319/. Accessed 20 Dec 2022.

Grigalashvili, V. (2022). E-government and E-governance: Various or multifarious concepts. *International Journal of Scientific and Management Research, 5*(1).

Harvard Business Review. (2022a). *How augmented reality can—And can't— Help your brand.* https://hbr.org/2022/03/how-augmented-reality-can-and-cant-help-your-brand. Accessed 6 Nov 2022.

Harvard Business Review. (2022b). *How the metaverse could change work.* https://hbr.org/2022/04/how-the-metaverse-could-change-work. Accessed 20 Dec 2022.

Hassani, H. (2022). A study of new emerged challenges for states to rule the cyberspace: Consequences of platformization and emergence of metaverse. *Political Science, 25*(98), 161–184.

Information Technology and Innovation Foundation. (2021). *Public policy for the metaverse: Key takeaways from the 2021 AR/VR Policy Conference.* https://itif.org/publications/2021/11/15/public-policy-metaverse-key-tak eaways-2021-arvr-policy-conference/. Accessed 20 Dec 2022.

Jena, S., Epari, V., & Sahoo, K. C. (2022). Integration of national cancer registry program with Ayushman Bharat Digital Mission in India: A necessity or an option. *Public Health in Practice, 3*, 100263.

Jung, Y. (2022). Current use cases, benefits and challenges of NFTs in the museum sector: Toward common pool model of NFT sharing for educational purposes. *Museum Management and Curatorship, 38*(4), 451–467.

Kanematsu, H., Kobayashi, T., Barry, D. M., Fukumura, Y., Dharmawansa, A., & Ogawa, N. (2014). Virtual STEM class for nuclear safety education in metaverse. *Procedia Computer Science, 35*, 1255–1261.

Kim, S. H., & Yoo, J. Y. (2021). A study on the recognition and acceptance of metaverse in the entertainment industry. *Journal of the Korea Entertainment Industry Association (JKEIA), 15*(7), 1.

Kshetri, N. (2022). Policy, ethical, social, and environmental considerations of Web3 and the metaverse. *IT Professional, 24*(3), 4–8.

Lee, H. J., & Gu, H. H. (2022). Empirical research on the metaverse user experience of digital natives. *Sustainability, 14*(22), 14747.

Lv, Z., Shang, W. L., & Guizani, M. (2022). Impact of digital twins and metaverse on cities: History, current situation, and application perspectives. *Applied Sciences, 12*(24), 12820.

Madou, M. (2022). Now is the time to strengthen cyber defences. *Network Security, 2022*(8).

Mohanty, L., & Swain, S. C. (2022). Use of digital technologies by the Msmes to preserve cultural heritage of India and achieve sustainable development goals. *ECS Transactions, 107*(1), 14343.

Mondaq. (2022). *India: Metaverse: Legality & regulatory concerns in India.* https://www.mondaq.com/india/fin-tech/1195182/metaverse-legality-reg ulatory-concerns-in-india. Accessed 19 Dec 2022.

Niti Ayog. (2022). *Meta-governance: Role of metaverse in India's e-Governance.* https://www.niti.gov.in/index.php/meta-governance-role-metaverse-indias-e-governance. Accessed 19 Dec 2022.

Niu, X., & Feng, W. (2022). Immersive entertainment environments—From theme parks to metaverse. In *International Conference on Human-Computer Interaction* (pp. 392–403). Springer.

People Matters. (2022). *How the metaverse can reshape the future of work.* https://www.peoplematters.in/article/training-development/how-the-metaverse-can-reshape-the-future-of-work-35652. Accessed 20 Dec 2022.

Pozniak, H. (2022). Could engineers work in the metaverse? *Engineering & Technology, 17*(4), 1–8.

PS Newswire. (2022). *Metaverse in entertainment market to reach $221.7 billion, globally, by 2031 at 32.3% CAGR: Allied market research.* https://www.prnewswire.com/news-releases/metaverse-in-entertainment-market-to-reach-221-7-billion-globally-by-2031-at-32-3-cagr-allied-market-research-301691 463.html. Accessed 20 Dec 2022.

Qi, P., & Chen, Z. (2022). The origin, characteristics and prospect of metaverse. *Advances in Education, Humanities and Social Science Research, 1*(1), 315–315.

Rospigliosi, P. A. (2022a). Adopting the metaverse for learning environments means more use of deep learning artificial intelligence: This presents challenges and problems. *Interactive Learning Environments, 30*(9), 1573–1576.

Rospigliosi, P. A. (2022b). Metaverse or simulacra? Roblox, Minecraft, Meta and the turn to virtual reality for education, socialisation and work. *Interactive Learning Environments, 30*(1), 1–3.

Stanoevska-Slabeva, K. (2022). Opportunities and challenges of metaverse for education: A literature review. *EDULEARN22 Proceedings*, 10401–10410

Theis, T. N., & Wong, H. S. P. (2017). The end of Moore's law: A new beginning for information technology. *Computing in Science & Engineering, 19*(2), 41–50.

Thomason, J. (2022). Metaverse, token economies, and non-communicable diseases. *Global Health Journal, 6*(3), 164–167.

Tong, A. (2022). Non-fungible token, market development, trading models, and impact in China. *Asian Business Review, 12*(1), 7–16.

Trade Finance Global. (2022). *Open sesame: Trade finance in the metaverse.* https://www.tradefinanceglobal.com/posts/open-sesame-trade-fin ance-in-the-metaverse/. Accessed 20 Dec 2022.

Upadhyay, A. K., & Khandelwal, K. (2022). Metaverse: The future of immersive training. *Strategic HR Review, 21*(3), 83–86.

Wang, G., Badal, A., Jia, X., Maltz, J. S., Mueller, K., Myers, K. J., Niu, C., Vannier, M., Yan, P., Yu, Z., & Zeng, R. (2022a). Development of metaverse for intelligent healthcare. *Nature Machine Intelligence, 4*(11), 922–929.

Wang, H., Chen, D., & Deng, Q. (2022b). The Formation, Development and Research Prospect of Educational Metaverse. *Education Journal, 11*(5), 260-266.

Wang, M., Yu, H., Bell, Z., & Chu, X. (2022c). Constructing an edu-metaverse ecosystem: A new and innovative framework. *IEEE Transactions on Learning Technologies.*

Wang, X., Wang, J., Wu, C., Xu, S., & Ma, W. (2022d). Engineering brain: Metaverse for future engineering. *AI in Civil Engineering, 1*(1), 1–18.

Wang, Y., Su, Z., Zhang, N., Xing, R., Liu, D., Luan, T. H., & Shen, X. (2022e). A survey on metaverse: Fundamentals, security, and privacy. *IEEE Communications Surveys & Tutorials.*

Xu, M., Ng, W. C., Lim, W. Y. B., Kang, J., Xiong, Z., Niyato, D., Yang, Q., Shen, X. S., & Miao, C. (2022a). A full dive into realizing the edge-enabled metaverse: Visions, enabling technologies, and challenges. *IEEE Communications Surveys & Tutorials, 25*(1), 656–700.

Xu, X., Zou, G., Chen, L., & Zhou, T. (2022b). Metaverse space ecological scene design based on multimedia digital technology. *Mobile Information Systems, 2022.*

CHAPTER 6

Way Forward For Metaverse Adoption

Abstract This is a concluding chapter that describes metaverse technology challenges posed on businesses as well as users. Ethical consideration is also a crucial point to discuss especially when the metaverse possesses difficulties such as cyberbullying, privacy and fairness and safety. Business implications for both consumers, as well as enterprises, discuss the opportunity landscape in creating end-to-end journeys immersive. In the end, the chapter highlights what can be done today to prepare for the future.

Keywords Security · Privacy · Fairness · Cyberbullying · Social issues · Safety · Transparency

6.1 Business Framework for Metaverse Adoption

The metaverse, a term once relegated to the pages of science fiction, is rapidly becoming a tangible reality. As we stand on the cusp of this digital frontier, it's imperative for different business leaders to understand its implications and the considerations needed for successful adoption within enterprises.

As shown in Fig. 6.1, there are certain key pointers to be considered by business leaders for Metaverse adoption.

R. Gupta and S. K. Pal, *Introduction to Metaverse*,
https://doi.org/10.1007/978-981-99-7397-2_6

Fig. 6.1 Key pointers for business leaders to consider while adopting Metaverse

i. **Strategic Alignment**: Before diving into the metaverse, leaders should ask themselves how it aligns with their company's mission, vision, and strategic goals. Is it a mere industry trend, or can it genuinely enhance value proposition.

ii. **Customer Experience**: The metaverse offers a unique, immersive customer experience. Business Leaders can consider how they can leverage it to deepen customer engagement, provide unparalleled service, or offer innovative products.

iii. **Employee Collaboration**: Beyond customer engagement, the metaverse can revolutionize how enterprise teams collaborate. Virtual workspaces can foster creativity, break down geographical barriers, and create a sense of unity in a dispersed workforce.

iv. **Technological Infrastructure**: The metaverse demands robust technological infrastructure. Assess enterprise's current capabilities and investing in areas like VR/AR hardware, high-speed connectivity, and cloud computing can ensure a seamless metaverse experience.

v. **Data Privacy and Security**: With great digital power comes great responsibility. Ensuring that metaverse endeavours prioritize user data protection, adhere to privacy regulations, and employ state-of-the-art security measures, will be key for the business leaders.

vi. **Ethical Considerations**: The metaverse presents ethical dilemmas, from digital inclusivity to the potential for creating addictive environments. It's crucial to approach these challenges with a strong ethical compass, ensuring that an enterprise's actions benefit both stakeholders and society at large.

vii. **Skill Development and Training**: The metaverse requires new skills. Investing in training of the business workforce in areas like

3D design, virtual event management, and digital asset creation can be valuable. Also, partnering with educational institutions or offering internal courses can help smooth internal adoption of the Metaverse.

viii. **Financial Implications**: Beyond the initial investment, consider the long-term financial implications. How will the metaverse drive revenue? What are the potential costs? A clear financial roadmap will ensure that an organization's metaverse initiatives are sustainable and profitable.

ix. **Regulatory Landscape**: The metaverse is a new domain, and regulatory frameworks are still evolving. Leaders will need to stay abreast of local and international regulations, ensuring that their enterprise remains compliant as it navigates this digital realm.

x. **Continuous Learning and Adaptation**: The metaverse is dynamic, with technologies, user preferences, and best practices evolving rapidly. Fostering a culture of continuous learning, encouraging teams to experiment, iterate, and adapt, will be really beneficial as part of upskilling and cross-skilling programmes.

Metaverse is not just another technology trend; it's a paradigm shift that has the potential to redefine industries and reshape how we do business. As leaders, their vision, foresight, and adaptability will determine how successfully their enterprise navigates this new dimension. Embrace the metaverse with an open mind, a strategic approach, and a commitment to creating value for all stakeholders.

Adopting metaverse technology is a significant decision for businesses, given its transformative potential. A structured decision-making framework can help businesses evaluate the opportunities and challenges of the metaverse and make informed choices. Here's a suggested framework:

---

1. **Define Clear Objectives**

- **Purpose**: Understand why you want to enter the metaverse. Is it for brand visibility, new revenue streams, customer engagement, or internal collaboration?
- **Value Proposition**: Determine what unique value you can offer in the metaverse. How will your presence benefit users or customers?

2. Conduct a SWOT Analysis

- **Strengths**: What are your business's strengths that can be leveraged in the metaverse? This could be brand reputation, technological expertise, or content creation capabilities.
- **Weaknesses**: Identify potential weaknesses or gaps in your current capabilities that could hinder metaverse adoption.
- **Opportunities**: Explore the potential opportunities in the metaverse, such as new customer segments, partnerships, or markets.
- **Threats**: Recognize potential threats, like competitors, technological changes, or regulatory challenges.

3. Evaluate Technological Readiness

- **Infrastructure**: Assess if your current technological infrastructure can support metaverse initiatives. This includes hardware, software, and network capabilities.
- **Skills and Expertise**: Determine if you have the necessary in-house expertise or if you need to hire or partner with experts in VR, AR, blockchain, or 3D graphics.

4. Financial Analysis

- **Cost Assessment**: Estimate the initial investment and ongoing costs of establishing and maintaining a presence in the metaverse.
- **Revenue Projections**: Forecast potential revenue streams from metaverse activities, such as virtual goods sales, virtual real estate, or services.
- **ROI Calculation**: Compare the projected revenues against the expected costs to determine the potential return on investment.

5. Consider Ethical and Social Implications

- **Digital Inclusion**: Ensure that your metaverse initiatives are inclusive and accessible to diverse user groups.
- **Privacy and Security**: Prioritize user data protection and ensure secure transactions within the metaverse.

6. Develop a Pilot Project

- **Scope**: Start with a small, well-defined project to test the waters. This could be a virtual showroom, a digital event, or an internal collaboration space.
- **Feedback Loop**: Gather feedback from users, stakeholders, and employees to understand what works and what doesn't.

7. **Scale and Integrate**

- **Integration**: If the pilot is successful, consider how metaverse initiatives can be integrated into your broader business strategy.
- **Expansion**: Explore opportunities to scale your presence, diversify offerings, or expand into new areas of the metaverse.

8. **Continuous Review and Adaptation**

- **Monitor Trends**: The metaverse is an evolving space. Stay updated with technological advancements, user preferences, and competitive moves.
- **Iterate**: Regularly review and refine your metaverse strategy based on feedback, performance metrics, and changing business goals.

Adopting metaverse technology is a strategic decision that requires careful consideration of various factors. This framework provides a structured approach to evaluate the potential of the metaverse for your business and make informed decisions. By following this framework, businesses can systematically approach the metaverse, ensuring that their adoption is aligned with their strategic goals and delivers tangible value.

6.2   CHALLENGES OF METAVERSE TECHNOLOGY

The possibilities and promises posed by metaverse are enormous and various companies are developing products, apps, and services for helping in the development of the metaverse and serving its users with a more immersive digital world. However, there is a challenging side to the metaverse.

6.2.1   Privacy Issues

As per Wang et al. (2022), the metaverse is the next version of the internet and it uses technology such as AR and VR to immerse people in the digital world. Privacy comes as a primary challenge. The technology that tracks behaviour online also exists in the metaverse and it will become more invasive and intense in the upcoming years.

VR sets tend to collect more data than smartphones. For instance, companies are able to connect extensive data biometrics, eye movements, and physiological responses. Some of these identifiers, like unique identifiers, are collected by technology companies to tailor user experiences based on personal preferences. However, this raises concerns as the data, if it falls into the wrong hands, can result in loss of personal and national wealth and reputation. It can expose the private information of the people to outsiders giving them access to intricate details that can be used in harming, individuals, companies, or a group of people (Falchuk et al., 2018). Thus, for a better experience, it is crucial for the creators of metaverse applications to enhance their privacy features. The metaverse should be a secure environment where both the rights of users and their data are protected.

6.2.2   *Fairness*

Various virtual worlds will be created in the metaverse and each virtual world may have its own set of laws to control user behaviour and activities. As a result, the required amount of time to manage and maintain such virtual worlds would be immense. It is crucial to note that autonomous systems in the virtual world rely on AI algorithms in order to respond to the dynamic changes in avatars and virtual items (Bibri & Allam, 2022). This emphasizes the relevance of user impressions of machine learning algorithm fairness, that is, perceived fairness. Collective, as well as individual results that are unfavourable to them, might be disastrous. Metaverse designers should create channels to collect the voices of various community groups and work together to create solutions that promote justice in the metaverse ecosystems.

6.2.3   *Cyberbullying*

As documented by Karanfiloğlu and Sağlam (2022), cyberbullying refers to misbehaviour through the web that includes uploading, sending, or spreading damaging, unpleasant, malicious, or false material in cyberspaces. Metaverse is considered to be massive cyberspace. As a result, cyberbullying in the metaverse might be an unavoidable societal danger to the ecology. Some large authorities propose that some virtual worlds in the metaverse be stopped working since the metaverse will not be able to function indefinitely.

6.2.4 Social Issues

Cultural acceptance of devices linking individuals to the metaverse, which refers to the public or bystanders' acceptance of such technologies as mobile AR and VR headsets, requires additional examination. Further, user protection of mobile headsets can negatively influence onlookers and users, resulting in the breakdown of users in the virtual world (Allam et al., 2022).

Furthermore, it is important to evaluate the user acceptability of avatars or digital copies of people, at distinct times. In addition to this, the metaverse can be viewed as a massive digital environment that will be powered by a vast number of computational resources. As a result, the metaverse has a lot of capabilities to use a lot of energy and pollute the digital environment. Considering that the metaverse must not deprive future generations, it is important for the developers to consider green computing while developing metaverse applications (Hutson, 2022). Environmental protection and accountability can influence user affection and views towards the metaverse and the number of active users as well as opponents. As a result, sourcing and creating metaverse used of data analytics based on sustainability indices will be a significant requirement for its wider adoption.

6.2.5 Accountability

According to Smaili and de Rancourt-Raymond (2022), accountability is crucial in realizing the full potential of the metaverse ecosystem. Despite advancements in technology making pervasive computing a reality, many benefits won't be fully realized unless people become familiar and accepting of the technology. Accountability is essential for building trust and it involves the responsibilities, methods, and incentives for those who implement, manage, design, and provide services in the metaverse. Another facet of accountability in the universe of the metaverse is how the data of the users is managed such as users' position and surroundings, instead of standard smart gadgets. The metaverse universe must encourage the data reduction principle for addressing this issue.

6.2.6   Safety

As Smaili and de Rancourt-Raymond (2022) point out, safety also plays a key role in realizing the full potential of the metaverse. With the merging of reality and virtual reality in the largely unsupervised and unregulated metaverse, the risk of harassment increases. Establishing early measures to protect users will not only benefit individuals and the platform, but also help to address the trust gap. Companies in the immersive technology field can be encouraged to develop applications that promote a safer environment in the metaverse, making it accessible to a broader range of users.

According to a study by Wang et al. (2022), the metaverse, while appealing as an entertainment platform for young audiences, poses significant safety concerns for them. Without proper supervision, children might encounter harmful situations in these expansive digital realms. A 2021 report from the Centre for Countering Digital Hate (CCDH) highlighted concerns regarding Facebook's VR metaverse application, revealing that young users are at a heightened risk of encountering bullying, coerced repetition of prejudiced remarks, and exposure to distressing scenarios, such as implied violence.

Anshari et al. (2022) also noted that adults aren't exempt from potential hazards in the metaverse, with some facing unsolicited explicit content and questions. Furthermore, communal areas within the metaverse have been observed to contain instances of discriminatory remarks and unwarranted virtual advances.

6.3   Preparedness for Metaverse Technology

6.3.1   Human Resource Preparedness

According to Allam et al. (2022), the metaverse has the potential to broaden the talent pool by breaking down geographical barriers in the workplace. The hiring process, including interviews and onboarding, can be done in a three-dimensional world, allowing new hires to form professional relationships and understand the company culture in an immersive environment. This can also provide HR departments with a way to evaluate potential candidates by testing their analytical skills in simulated scenarios. For this purpose, a streamlined metaverse-friendly recruitment and selection process needs to be prepared for its effective implementation.

As per the views of Zvarikova et al. (2022), the ability to concentrate on values, resource allocation, purpose, and culture is also required. Strategic HR is all about planning for the future of the workforce, such as planning, scheduling, and workforce management. In order words, who will be where at a given point in time. As technologists and the next generation is creating and designing the future of the workplace, HR must be onboard to think about the innovation and vision in tandem. It is important for the metaverse to comprehend what the metaverse is and how it may assist the overall functions of HR.

Metaverse has immense potential in the learning and development aspects of the business. Computer-based training has been around since the 1990s. Annually, companies tend to invest a lot of funds into executive training programmes which were only accessible to a certain group of people (Hawkins, 2022b). This can be further expanded through simulation-based training in a 3D experience platform. This will further offer increased engagement, facilitating self-based learning, motivation and offering a memorable experience to the employees. However, human resource preparedness is required in the area of workplace conflicts that might get intensified will higher employee engagement in the work culture processes.

6.3.2 Security and Privacy

According to Su et al. (2022), the dream of a smooth and uninterrupted experience in the metaverse is becoming a reality every year. It is no surprise that the metaverse is an attractive target for cybercriminals who see a new opportunity to gain access to personal information, virtual assets, and digital currency. With the increasing popularity of the metaverse and the rise of Web 3.0, users are exposed to greater risks. Brands will be eager to interact with their users in innovative ways, but the more interaction points there are, the more opportunities there are for fraudsters to exploit. Therefore, it is inevitable that there is a requirement of a different approach to security in the virtual world as compared to the physical world.

Considering the current state of laws and provisions, the world is still not ready to deal with the flaws that the metaverse brings with it (Far & Rad, 2022). According to privacy advocates, it is crucial to establish consistent privacy regulations and standards, along with proper enforcement of these standards. Policymakers will need to address questions about how much monitoring of users is acceptable in the virtual world, what information can be tracked, and how much third-party content can be made available to users. Clear laws regarding data collection, storage, and usage will have to be established. Quick and effective action must be taken to address any privacy violations (Buck & McDonnell, 2022).

As per Cao (2022), another concern that has been raised is the possibility of illegal mass surveillance. With the ability to access a large amount of citizen data, governments may be able to constantly monitor individuals. In such cases, the adoption of an updated user consent has to be initiated. Keeping it in simple language and regularly updating the same can help individuals in restricting personal data to be available on the wider platform. Therefore, it is important to establish clear guidelines for obtaining user consent, to ensure that individuals have control over what personal information and personas are linked to their real-world identity.

6.3.3   Threat to Humans

In the obvious sense, the metaverse is an iteration of the internet and gives far more immersive experiences to people. However, it brings a significant threat to humans along with the long range of facilities it offers to make the daily processes technologically friendly (Oleksy et al., 2022). Higher accessibility to the data has raised the question of the theft of identity. Recent research says that identity theft damages were estimated to be USD 24 billion.

Personal information is also collected by legitimate businesses. On the other hand, virtual reality has the potential to push data to a level that may be out of reach in the present scenario. For instance, virtual reality headsets can hypothetically allow third parties to collect more personal and sensitive information in the form of biometric data, voiceprint data, and even facial geometry (Buana, 2023).

Ransome is another malicious software that encrypts the data and prevents people from accessing them. It then asks people to pay a particular amount of money to gain access to their own data. Thus, it has been termed "Ransomware" (Dunnett et al., 2022). When it comes to the metaverse, it will have much more information than a regular social media page, including a lot of sensitive information which threatens the existence of humans and their monetary and non-monetary belongings.

The metaverse is built on the idea of bringing people closer and together despite their geographical limitations. Though it might be beneficial in certain aspects, it can also raise issues. In the metaverse, one has to deal with people who have contrary viewpoints. As per Gorichanaz (2022), people tend to behave differently in the virtual world as compared to the real world. This is particularly pronounced in the context of massively multiplayer online role-playing games, where experienced players often exhibit hostile behaviour towards new players and engage in sexual harassment towards female players.

6.4   Potential Misuse of Metaverse Technology

The potential misuse of metaverse technology is plentiful ranging from unlimited data harvesting impacting privacy, harassment, constant abuse, and imposter avatars stealing sensitive information. It is expected that cybercriminal activities will grow to $10.5 trillion in 2025 (The Digital Speaker, 2022). The amount is more than the combined GDP of France, Germany and the UK. Also, it is more than the entire global eCommerce or commercial real estate industry.

The anonymity and privacy provided by cryptocurrencies and the metaverse can create opportunities for illegal activities, including child sexual abuse and other forms of exploitation. The challenge for governments and law enforcement agencies will be to develop and enforce regulations that prevent such activities without stifling innovation or restricting the benefits of the metaverse. It will also be essential to establish safeguards and mechanisms to protect individuals from harassment and abuse within the metaverse. This will require a multi-stakeholder approach, involving policymakers, technology companies, and civil society organizations.

As per Anshari et al. (2022), in the metaverse, it is quite possible for an individual to enter into someone's personal space without being later aware of who the former is. Given the multisensory capabilities of the metaverse, inclusive of the haptic technology, that is the sense of touch, the impact and experience might far worsen. Arguably, the metaverse presents itself as a tool that can easily be misused for sexual abuse, bullying, and intimidation. Indeed, there have been recent media reports that some VR-based games that are accessible to children contain certain inappropriate content (Baker-Brunnbauer, 2022). Additionally, the use of AI-generated deepfakes has the potential to spread false information that is indistinguishable from reality. This can have significant implications in spreading misleading and potentially harmful content, especially when it comes to political figures and inflammatory statements. The already prevalent issue of misinformation on platforms such as WhatsApp would only be exacerbated by the use of deepfakes.

6.5   ETHICAL CONSIDERATIONS FOR METAVERSE

While the law often aligns with widely accepted ethical principles, there may be instances where legal requirements do not fully reflect what is considered to be morally right (Kshetri, 2022). These ethical standards and morals are not bound by any country or borders. People from different countries, ethnic groups or regions might hold similar value systems and morals. Since the metaverse shows its existence at the global level, ethics and morality also need standardization.

Considering the human and ethical side of the metaverse, one of the first problems that are encountered is the reduced physical connection and disconnection from the real world. With the highly engaging and immersive experience offered by the metaverse, the chances of lost physical connection are at a surge. As per Han et al. (2022), new technologies have always demanded societal conversations and the metaverse is no different with its ethical implication being put into question now more than ever.

As per the Gartner (2022e), according to predictions, by 2026, a quarter of the population will spend at least an hour in the metaverse. Artificial Intelligence, a vital component of the metaverse, can solve many complex problems but it is not without flaws. AI systems can be biased due to the biased data fed into them, creating an ethical dilemma. To combat this, transparency and accountability in AI are crucial, along with efforts to ensure that the AI is free from bias. For example, Meta recently launched initiatives to provide free education in technology, with a focus on increasing the representation of racial minorities in the field, in an attempt to create AI systems that are free from bias.

By 2026, it is expected that over a quarter of people will spend at least one hour in the metaverse. Although AI plays a crucial role in the metaverse, it has its flaws, such as the potential for bias to be introduced into the system through prejudiced data inputs. To ensure an ethical and inclusive metaverse, it is essential to address the digital divide by reducing existing inequalities and promoting diversity in the tech industry (Salmasi & Gillam, 2009). Currently, only a small percentage of tech jobs globally are held by people of colour and women. Bridging the gap requires a change in attitudes and actions, and it is crucial to make underrepresented groups feel welcome in the metaverse.

The last 10 years has experienced the emergence of digital assets and blockchain technology which aims to decentralize communities, and democratize access to key goods, services, and experiences. According to Golf-Papez et al. (2022), if the majority of shares of the metaverse is owned operated, governed, or owned by the tech giants, it will give rise to the problems, such as privacy, manipulations, theft, and conditioning to favour some. For instance, if Meta or Amazon controls the functions of the metaverse, it is only natural that their business interest will take over. It will endure favouritism in the market and distort ethical standards. It is important to democratize the platform. Having a decentralized environment which is not governed by one large entity will ensure higher transparency and the run of a safe algorithm (Yin et al., 2022).

The internet and social media have already reflected that when behind the protection of phone screens and computers, people tend to vocalize their opinion that they might never express in the real life. Such platforms have become the breeding ground of bullies, toxicity, and hate speech. Yet the law and enforcement appear to lag behind in the process of technological development. It is only logical to say that the same ill-mannered behaviour will be expressed in the metaverse, where people can easily hide behind their digital avatar or digital twin. An ethical framework is demanded around this area for better dealing with such practices that are morally unethical.

6.6   BUSINESS IMPLICATIONS

6.6.1   For Consumers

According to Dwivedi et al. (2022b), the metaverse platform can profoundly change how consumers and businesses interact with products, and services as well as each other. For instance, consumers can easily hop from one competing for the virtual car dealership to another, with the immersive feeling of the wind in their hair as they test drive. Today, many young consumers are already using the facility of virtual try-on of clothes at virtual retail stores or buying virtual merchandise for their virtual gaming environment. Metaverse is growing at a remarkable pace by blending the new form of interactivity that emulsifies the digital and physical world and rewriting the rules of the consumer economy. The new consumer landscape in the metaverse will help in creating end-to-end customer journeys, open spaces for communities and immersive experiences to accelerate creativity. It will contribute in increasing revenues as well as optimising cost of human figures.

According to the EY Future Consumer Index, 10% of consumers are already involved with the metaverse through activities such as using crypto-wallets, purchasing virtual goods, or experimenting with augmented and virtual reality. Virtual environments and ecosystems provide consumers with an imaginative extension of reality, where they can unwind, build communities, connect with others, solve problems, experience novel forms of brand value, and explore alternative realities. The gaming industry and immersive ecosystems present numerous chances for consumers to discover the broad range of opportunities that companies have to offer them.

Consumers are also showing interest in the metaverse by attending virtual fashion shows and live concerts (Rospigliosi, 2022b). Apart from this, consumers are also showing interest in e-sports and massively multiplayer online games are attending new audience that behaves as communities in a novel, interoperable, and inclusive environment. It will improve the efficiency by better management of the crowd and enhance revenues by accommodating larger crowd without any physical investment.

Hence, consumers are expected to have better experience of products and services which are possible in virtual world. Many of the digital twins will be available to the consumers with the help of Metaverse technology giving them easy access at an affordable cost.

6.6.2   For Enterprises

With consumers investing more in their online persona, they are purchasing more digital goods and devoting more time to virtual experiences. Their digital identity has become just as significant as their physical identity, blending the digital and physical worlds together. As consumers are spending more time in the digital realm, retailers and brands must also have a presence there. The metaverse provides exceptional chances to craft one-of-a-kind and distinctive experiences to connect with consumers and establish brand excitement, loyalty, and trust. Failing to do so could lead to becoming irrelevant (Schmitt, 2022).

As the metaverse continues to evolve, brands have the opportunity to shape its offerings. According to Barrera and Shah (2023), the metaverse can streamline the process of bringing a product to market by allowing for various stages, such as trial production testing, marketing, operations management, and others, to be tested and verified within the virtual community. This could significantly shorten the time it takes to bring a product to market, compared to the traditional process that involves procurement of raw materials, prototyping, and testing phases.

As per Hotaran et al. (2022), shopping is not just the activity of purchasing a product, it has expanded to become an important experience of buying from a particular brand that resonates with consumers. With Q-commerce and E-commerce on the rise, the metaverse will be at the forefront of creating meaningful and innovative interactions. As an AR platform, it will help enterprises in offering a unique experience to the customer base. Additionally, according to Hung (2022), the popularity and visible growth of virtual characters in the retail sectors like fashion and gaming, has pushed the brands to enhance their digital experience for information dissemination and loyalty-building process.

The pandemic has encapsulated enterprises into remote work. Most companies are still continuing to work in the remote working module or hybrid work module. As noted by Damar (2021), the metaverse has the potential to enhance the work culture while still accommodating remote or hybrid work arrangements. Research has indicated that remote work can negatively impact collaboration among colleagues, due to the lack of face-to-face communication. The metaverse, with its persistent virtual reality workplace settings, can alleviate this issue by fostering informal conversations, interactions among employees, and teamwork.

According to Dwivedi et al. (2022b), the metaverse can enhance customer engagement and marketing efforts. Modern marketing and customer interactions have evolved significantly from traditional print ads, and now it's all about connecting with customers by allowing them to experience your world. The metaverse presents a unique opportunity for brands to communicate with their audience in a personalized way and elevate the shopping experience. Businesses will need to adapt their strategies to analyse digital insights, traffic, and virtual avatar purchasing patterns to reach their target audience effectively. This requires innovative approaches, refined SEO, and effective advertisement statistics analysis.

6.6.3   *Financial Implications*

The financial implications of the metaverse are vast and multifaceted. Let's delve into how the metaverse might reshape the financial landscape and what it means for businesses and individuals alike.

a. Creation of New Economic Systems

Digital Currencies and Cryptocurrencies: The metaverse is likely to have its own economic systems, often built on digital currencies or cryptocurrencies. These currencies can be used for transactions within the metaverse, from buying virtual goods to investing in virtual real estate.

Decentralized Finance (DeFi): DeFi platforms, which aim to recreate traditional financial systems (like lending and borrowing) without intermediaries, could find a natural home in the metaverse, allowing users to manage their finances in more decentralized ways.

b. Digital Assets and Ownership

Non-Fungible Tokens (NFTs): These are unique digital assets verified using blockchain technology. In the metaverse, NFTs can represent anything from virtual real estate to digital art or even a piece of music. Their uniqueness and verifiable ownership can make them valuable, leading to new investment opportunities.

Virtual Real Estate: As mentioned earlier, spaces within the metaverse can be bought and sold. As demand grows, these virtual properties might appreciate in value, similar to real-world real estate.

c. Investment Opportunities and Risks

Start-ups and Venture Capital: As the metaverse expands, there will be a surge in start-ups building tools, platforms, and experiences for this new digital frontier. This presents opportunities for venture capitalists and individual investors.

Speculative Investments: Like any emerging technology, there's a risk of speculative bubbles. Prices of virtual goods, real estate, or native metaverse currencies could be volatile, leading to potential financial booms or busts.

d. Evolution of Traditional Financial Institutions

Banks and Financial Services: Traditional banks might establish virtual branches or offer financial products tailored for the metaverse. This could range from virtual savings accounts to loans for buying virtual properties.

Insurance: New types of insurance products might emerge, covering risks associated with virtual assets, digital identity theft, or even virtual events.

e. Economic Impact on Traditional Industries

Entertainment and Media: The revenue models for entertainment might shift, with artists and creators monetizing directly through virtual concerts, shows, or selling digital merchandise.

Retail: Brands might see a shift in revenue from physical goods to digital counterparts. Virtual fashion or digital accessories for avatars could become significant revenue streams.

f. Regulatory and Tax Implications

Digital Transactions: Governments and regulatory bodies will need to figure out how to tax transactions in the metaverse, especially as the lines blur between virtual and real-world value.

Asset Valuation: Determining the value of digital assets, especially for tax purposes, will be a challenge. Is virtual real estate a taxable asset? How do you value a unique piece of digital art?

The financial implications of the metaverse are profound, reshaping how we think about value, ownership, and economic interaction. While it offers immense opportunities, it also comes with risks and uncertainties. Businesses, investors, and individuals must navigate this new financial landscape with both enthusiasm and caution. The metaverse, in essence, is set to redefine the very fabric of our financial systems, creating a blend of the physical and digital economies.

6.7   PREPARING FOR THE FUTURE: WHAT CAN BE DONE TODAY

In order to effectively navigate the metaverse, companies will need to embrace a flexible approach and embrace a digital transformation plan that is forward-thinking and will open up new opportunities for them. It is often said that challenges are beautiful opportunities in disguise (Lee, 2021). This completely fits the Metaverse, as many challenges are providing a lot of opportunities ahead for companies and customers.

Scheiding (2022), questioned where is Metaverse heading? Consumer, Businesses, finance, and public sectors all can gain from the many opportunities to be offered by Metaverse. To exploit all the opportunities, it is essential to understand how to prepare and what is required to get started. Enthusiasts of the Metaverse world say that the fast expansion of this technology is going to benefit every feature of society, be it education, health care, entertainment, gaming, and social. According to them, the infusion of more data in people's experiences, the development of AI systems and the building of new spaces for tech users could enhance their lives (Collins, 2008). Of course, there are concerns with all this digitization about safety, security, health, economic implications, and privacy.

A large number of technology experts were surveyed by Pew Research Center and Elon University's Imagining The Internet Center to gather their thoughts on the future of the metaverse. Out of the 624 experts who responded, 54% predicted that by 2040, the metaverse will be a seamless and fully immersive part of daily life for at least half a billion people worldwide (Pew Research Center, 2022b).

It is difficult to predict the future, yet companies have developed tools and strategies which help in predicting the trajectory of the technology and help strategists and companies to steer their organizations on the same path to maximize use of opportunities while mitigating the risks involved.

One of the popular prediction tools is the technology hype cycle by Gartner. It is helpful in predicting the trajectory of technology trends (Gartner, 2022e).

Elementally, the hype cycle illustrates five stages of any emerging technology.

- **Innovation Trigger**: Discovery of new technology.

- **The peak of Inflated expectation**: When the technology is in a transformational hyped phase without any clear evidence of its actual value (Damar, 2021).
- **Trough of Disillusionment**: where news and events show the negatives of technology as it is unable to deliver on its expectations.
- **The slope of Enlightenment**: When all the negative news subsides and technology starts delivering on its expectations.
- **Plateau Of Productivity**: Phase where technology finally generates value, subsides in the background, and becomes the commodity.

Using this technology hype cycle, it is possible to pin down the estimated positioning of the Metaverse (Fig. 6.2). The difficulty with this type of estimation is that to use this estimate it has to be considered that the Metaverse is a single technology (Oh & Kim, 2021). Right now, Metaverse can be seen as a group of technologies and companies and is projected as a platform such as computers and mobile phones. As a kind of platform, Metaverse can be segregated into three layers:

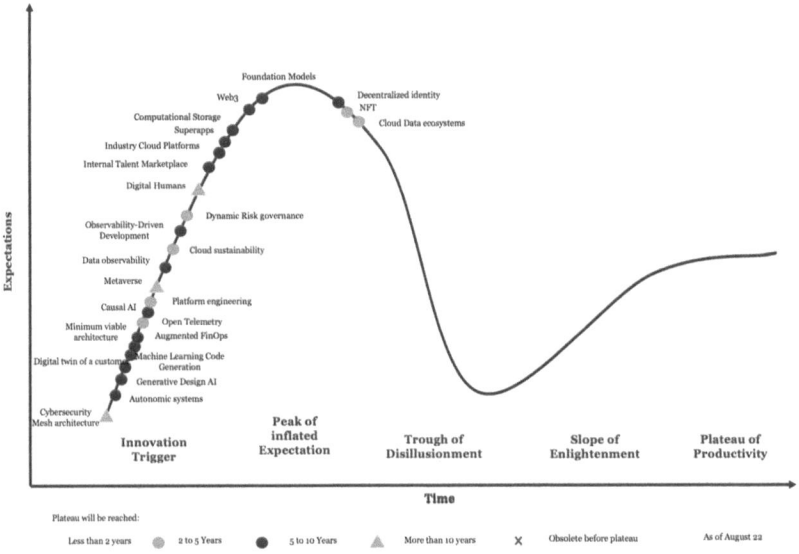

Fig. 6.2 Hype cycle for emerging technologies by Gartner (*Source* Gartner [2022e])

Operating system or backend: Systems where applications can be created and applied/deployed. This is the same as smartphone OSes—Android and iOS (two major mobile operating systems). Tech giants like Google, Apple, and Microsoft provide Metaverse operating systems.

Applications: This is the application layer itself that is made to work on operating systems. These are similar to the apps that are being downloaded on smartphones.

End-User Devices: These are the devices that allow users to experience innovative and enchanting experiences. Though Metaverse applications and OS are very complicated components still they offer quite a stable base to the whole experience. The major difficulty is the lack of interoperability between two layers which can block its way to being adopted as technology by the masses (Forbes, 2022d).

Microsoft holds a strong position in the personal computer operating systems and other tech companies like Google and Apple hold the dominant position in the smartphone operating systems. The important factor for the success of these systems is interoperability, which will help users effortlessly exchange data from computer to the smartphone (Phakamach et al., 2022).

Apart from the interoperability of systems, another major blocker in the success of Metaverse is the strong, affordable, and easy-to-use end-user devices. Devices should be advanced and easy to handle enough for the user to be consistent and for daily engagement. Once devices are capable of clicking, editing, and sharing photos with the same device, it is going to be one of the inflexion points for the Metaverse technology. AR and VR technology is already evolving and changing according to the businesses but still, they are nowhere close to the mass adoption and transformation power of the mobiles (Wang et al., 2022).

According to Yang et al. (2022), the metaverse has the potential to be as transformational as smartphones. Metaverse is somewhere between the "peak of inflated expectation" phase and "trough of disillusionment" stage. As of now, it does not carry the potential to be as immersive and transformational as others like electricity, the internet, smartphones, etc. Possibly there is a gap of a decade before it reaches the transformational phase but it is likely to reach there and businesses need to be prepared for it.

The very first step to be prepared for Metaverse is businesses have to identify, understand, and decide in which of the three roles they will fit—the expert experimenter, the contributor or the activator, as also suggested by The Entrepreneur (2022). The next step after deciding the role the business fits in, it needs to strategies and make a strong online presence. To do this, they need to hire people who will prepare the businesses to enact themselves in the Metaverse.

Also, businesses should focus on customer experience. What kind of experience do they want their users to have with their product in Metaverse? If they are able to think a few steps ahead of what their customers want then it will help them to be prepared and wait for their customers to experience that when they emerge in Metaverse (Gao et al., 2022). Finally, it is important for businesses to be flexible and incorporate the new learnings about Metaverse as it unfolds. Being flexible and acceptive of any change will help businesses to always be ahead and prepared to meet users' expectations in Metaverse.

In short, the following checklist can be numbered on what can be done today to prepare for the future of the metaverse:

- Streamlining the workforce through metaverse-friendly recruitment and selection
- Comprehending overall functions of HR according to the concept of Metaverse
- Simulation-based training
- Bringing digital safety in place
- Introducing provisions guiding safety and security
- Bring operating systems and backend systems in place
- Improve end-user experience on devices
- Provide end-to-end encryption of data
- Provide end-to-end customer service
- Be flexible and incorporate continuous learning

6.8   CONCLUDING REMARKS

In summary, while the metaverse presents immense opportunities for social, cultural, and economic development, it also poses significant risks to individuals and society as a whole. To ensure that the metaverse evolves in a responsible and ethical manner, a comprehensive approach that involves legislation, regulation, responsible content creation and distribution, and incentivizing positive behaviour is essential. Moreover, global coordination and collaboration among nations and regional blocs, including the UN, will be critical to ensuring that the metaverse remains a safe and secure environment for all. As the metaverse continues to evolve, it is vital that we remain vigilant and proactive in our efforts to mitigate potential risks and maximize its potential benefits.

To ensure safety and security in the metaverse, relying solely on legislation would not be enough. Responsibility should also be placed on content creators and platforms that provide access to the virtual world. Therefore, it is crucial to design the metaverse infrastructure with safety measures in place from the start. We can learn from the issues faced by social media platforms, such as cyberbullying, fraud, and disinformation, and take steps to prevent these problems before they become widespread. This is important because changing behaviour after people have already grown accustomed to it can be challenging, and some damage may have already been done. Encouraging good behaviour in the metaverse may also require incentivization (Hollensen et al., 2022).

The metaverse is expected to revolutionize in the upcoming decade and the trajectory of growth cannot be predicted in advance. Therefore, it needs constant vigilance and global actions to be taken in a concerted manner. According to Mourtzis, Panopoulos, Angelopoulos, Wang and Wang (2022), the UN system is generally considered the primary keeper of international order. In the past two decades, various events have agonisingly made it evident that UN architecture requires a major overhaul. Regional political/economic blocs must be supported to ensure that all their members duly comply to regulations of metaverse.

In the public sector, the metaverse might look like an amorphous and futuristic mass of technologies at the first sight. It is hard to turn them into a meaningful outcome (Lee, 2021). For this purpose, the first step is to build understanding and knowledge and to identify how metaverse capabilities can help in providing value to public safety organizations. The second is to create a vision and identify the solutions and partners that can deliver it. The third is to implement and adapt, by adopting the technology to address public safety needs and then experimenting and refining it as part of this process.

Therefore, to avoid being left behind, it is crucial for public agencies to engage with it, understand it, and build strengthening capabilities into their strategies and operating models.

REFERENCES

Allam, Z., Sharifi, A., Bibri, S. E., Jones, D. S., & Krogstie, J. (2022). The metaverse as a virtual form of smart cities: Opportunities and challenges for environmental, economic, and social sustainability in urban futures. *Smart Cities, 5*(3), 771–801.

Anshari, M., Syafrudin, M., Fitriyani, N. L., & Razzaq, A. (2022). Ethical responsibility and sustainability (ERS) development in a Metaverse business model. *Sustainability, 14*(23), 15805.

Baker-Brunnbauer, J. (2022). Ethical challenges for the Metaverse development.

Barrera, K. G., & Shah, D. (2023). Marketing in the Metaverse: Conceptual understanding, framework, and research agenda. *Journal of Business Research, 155*, 113420.

Bibri, S. E., & Allam, Z. (2022). The Metaverse as a virtual form of data-driven smart cities: The ethics of the hyper-connectivity, datafication, algorithmization, and platformization of urban society. *Computational Urban Science, 2*(1), 1–22.

Buana, I. M. W. (2023). Metaverse: Threat or opportunity for our social world? In understanding Metaverse on sociological context. *Journal of Metaverse, 3*(1), 28–33.

Buck, L., & McDonnell, R. (2022). Security and privacy in the Metaverse: The threat of the digital human.

Cao, L. (2022). Decentralized AI: Edge intelligence and Smart Blockchain, Metaverse, Web3, and DeSci. *IEEE Intelligent Systems, 37*(3), 6–19.

Collins, C. (2008). Looking to the future: Higher education in the Metaverse. *Educause Review, 43*(5), 50–52.

Damar, M. (2021). Metaverse shape of your life for future: A bibliometric snapshot. *Journal of Metaverse, 1*(1), 1–8.

Dunnett, K., Pal, S., Jadidi, Z., & Jurdak, R. (2022). The role of cyber threat intelligence sharing in the Metaverse. *IEEE Internet of Things Magazine.*

Dwivedi, Y. K., Hughes, L., Baabdullah, A. M., Ribeiro-Navarrete, S., Giannakis, M., Al-Debei, M. M., ... & Wamba, S. F. (2022a). Metaverse beyond the hype: Multidisciplinary perspectives on emerging challenges, opportunities, and agenda for research, practice and policy. *International Journal of Information Management, 66*, 102542.

Dwivedi, Y. K., Hughes, L., Wang, Y., Alalwan, A. A., Ahn, S. J., Balakrishnan, J., ... & Wirtz, J. (2022b). Metaverse marketing: How the Metaverse will shape the future of consumer research and practice. *Psychology & Marketing, 40*(4), 750–776.

Entrepreneur. (2022). Introducing 'Touch' in the Metaverse. https://www.entrepreneur.com/en-in/technology/introducing-touch-in-the-metaverse/430606. Accessed 6 November 2022.

Falchuk, B., Loeb, S., & Neff, R. (2018). The social Metaverse: Battle for privacy. *IEEE Technology and Society Magazine, 37*(2), 52–61.

Far, S. B., & Rad, A. I. (2022). Applying digital twins in Metaverse: User interface, security and privacy challenges. *Journal of Metaverse, 2*(1), 8–16.

Forbes. (2022a). 6 top Metaverse coins. https://www.forbes.com/advisor/inv esting/cryptocurrency/top-metaverse-coins/. Accessed 20 November 2022.

Forbes. (2022b). A short history of the Metaverse. https://www.forbes.com/ sites/bernardmarr/2022/03/21/a-short-history-of-the-metaverse/?sh=6e6 4a9ad5968. Accessed 6 November 2022.

Forbes. (2022c). Disney: The Metaverse, digital transformation, and the future of storytelling. https://www.forbes.com/sites/bernardmarr/2022/10/07/ disney-the-metaverse-digital-transformation-and-the-future-of-storytelling/? sh=19849efc13c0. Accessed 21 November 2021.

Forbes. (2022d). Meta's VR vs Apple's AR strategy-who will ultimately win? https://www.forbes.com/sites/timbajarin/2022/10/11/metas-vr-vs-app les-ar-strategy-who-will-ultimately-win/?sh=7ae676bb44ed.   Accessed   21 November 2022.

Gao, Y., Lu, Y., & Zhu, X. (2022). Metaverse, the future materials science computation platform based on metaverse. *The Journal of Physical Chemistry Letters, 14*, 148–157.

Gartner. (2022a). Gartner predicts 25% of people will spend at least one hour per day in the Metaverse by 2026. Press Release. https://www.gartner.com/ en/newsroom/press-releases/2022-02-07-gartner-predicts-25-percent-of-peo ple-will-spend-at-least-one-hour-per-day-in-the-metaverse-by-2026. Accessed 6 October 2022.

Gartner. (2022b). What is a Metaverse? And should you be buying in? Information Technology. https://www.gartner.com/en/articles/what-is-a-metaverse. Accessed 17 October 2022.

Gartner. (2022c). Metaverse evolution will be phased; Here's what it means for tech product strategy. https://www.gartner.com/en/articles/metave rse-evolution-will-be-phased-here-s-what-it-means-for-tech-product-strategy. Accessed 6 October 2022.

Gartner. (2022d). Gartner predicts 25% of people will spend at least one hour per day in the Metaverse by 2026. https://www.gartner.com/en/ newsroom/press-releases/2022-02-07-gartner-predicts-25-percent-of-people-will-spend-at-least-one-hour-per-day-in-the-metaverse-by-2026. Accessed 4 January 2022.

Gartner. (2022e). Gartner predicts 90% of current enterprise blockchain platform implementations will require replacement by 2021. Newsroom. https://www.gartner.com/en/newsroom/press-releases/2019-07-03-gartner-predicts-90--of-current-enterprise-blockchain. Accessed 8 October 2022.

Gartner. (2022f). What is new in the 2022 Gartner Hype Cycle for emerging technologies. https://www.gartner.co.uk/en/articles/whats-new-in-the-2022-gartner-hype-cycle-for-emerging-technologies. Accessed 3 January 2022.

Golf-Papez, M., Heller, J., Hilken, T., Chylinski, M., de Ruyter, K., Keeling, D. I., & Mahr, D. (2022). Embracing falsity through the Metaverse: The case of synthetic customer experiences. *Business Horizons, 65*(6), 739–749.

Gorichanaz, T. (2022). Being at home in the Metaverse? Prospectus for a social imaginary. *AI and Ethics*, 1–12.

Han, D. I. D., Bergs, Y., & Moorhouse, N. (2022). Virtual reality consumer experience escapes: Preparing for the Metaverse. *Virtual Reality*, 1–16.

Hawkins, M. (2022a). Metaverse live shopping analytics: Retail data measurement tools, computer vision and deep learning algorithms, and decision intelligence and modeling. *Journal of Self-Governance & Management Economics, 10*(2).

Hawkins, M. (2022b). Virtual employee training and skill development, workplace technologies, and deep learning computer vision algorithms in the immersive Metaverse environment. *Psychosociological Issues in Human Resource Management, 10*(1).

Hawkins, M. (2022c). Virtual employee training and skill development, workplace technologies, and deep learning computer vision algorithms in the immersive Metaverse environment. *Psychosociological Issues in Human Resource Management, 10*(1), 106–120.

Hollensen, S., Kotler, P., & Opresnik, M. O. (2022). Metaverse—The new marketing universe. *Journal of Business Strategy, 44*(3), 119–125.

Hotaran, I., Poleac, D., & Vrana, N. (2022). Five steps for sustainable business modelling in the Metaverse. *Fostering Recovery through Metaverse Business Modelling*, 530.

Hung, H. T. B. (2022). Keep your eyes on China's Metaverse: Another tool for maintaining its national security. *The Journal of Intelligence, Conflict, and Warfare, 5*(2), 1–31.

Hutson, J. (2022). Social virtual reality: Neurodivergence and inclusivity in the Metaverse. *Societies, 12*(4), 102.

Karanfiloğlu, M., & Sağlam, M. (2022). Media literacy, fact-checking and cyberbullying. *Organized by*, 37.

Kshetri, N. (2022). Policy, ethical, social, and environmental considerations of Web3 and the Metaverse. *IT Professional, 24*(3), 4–8.

Lee, B. K. (2021). The Metaverse world and our future. *Review of Korea Contents Association, 19*(1), 13–17.

Mourtzis, D., Panopoulos, N., Angelopoulos, J., Wang, B., & Wang, L. (2022). Human centric platforms for personalized value creation in Metaverse. *Journal of Manufacturing Systems, 65*, 653–659.

Oh, M. J., & Kim, J. (2021). An essay on the future of Metaverse as the harmony space both of Homo Ludens and Homo Fabre. *Journal of the Korea Convergence Society, 12*(9), 127–136.

Oleksy, T., Wnuk, A., & Piskorska, M. (2022). Migration to the Metaverse and its predictors: Attachment to virtual places and Metaverse-related threat. *Computers in Human Behavior,* 107642.

Pew Research Center. (2022a). The Metaverse in 2040. https://www.pewresearch.org/internet/2022/06/30/the-metaverse-in-2040/. Accessed 3 January 2022.

Pew Research Center. (2022b). About three-in-ten U.S. adults say they are 'almost constantly' online. https://www.pewresearch.org/fact-tank/2021/03/26/about-three-in-ten-u-s-adults-say-they-are-almost-constantly-online/. [Accessed 6 November 2022].

Phakamach, P., Senarith, P., & Wachirawongpaisarn, S. (2022). The Metaverse in education: The future of immersive teaching & learning. *RICE Journal of Creative Entrepreneurship and Management, 3*(2), 75–88.

Rospigliosi, P. A. (2022a). Adopting the Metaverse for learning environments means more use of deep learning artificial intelligence: This presents challenges and problems. *Interactive Learning Environments, 30*(9), 1573–1576.

Rospigliosi, P. A. (2022b). Metaverse or Simulacra? Roblox, Minecraft, Meta and the turn to virtual reality for education, socialisation and work. *Interactive Learning Environments, 30*(1), 1–3.

Salmasi, A. V., & Gillam, L. (2009). Machine ethics for gambling in the Metaverse: An. *Journal For Virtual Worlds Research, 2*(3), 1–23.

Scheiding, R. (2022). Designing the future? The Metaverse, NFTs, & the future as defined by unity users. *Games and Culture,* 15554120221139218.

Schmitt, M. (2022, July 21). Metaverse: Bibliometric review, building blocks, and implications for business, government, and society. *Building Blocks, and Implications for Business, Government, and Society.*

Smaili, N., & de Rancourt-Raymond, A. (2022). Metaverse: Welcome to the new fraud marketplace. *Journal of Financial Crime,* (ahead-of-print).

Su, Z., Zhang, N., Liu, D., Luan, T. H., & Shen, X. (2022). A survey on Metaverse: Fundamentals, security, and privacy.

The Digital Speaker. (2022). What are the dangers of the Metaverse? https://www.thedigitalspeaker.com/what-are-the-dangers-of-the-metaverse/. Accessed 4 January 2022.

Wang, F. Y. (2022). Metavehicles in the Metaverse: Moving to a new phase for intelligent vehicles and smart mobility. *IEEE Transactions on Intelligent Vehicles, 7*(1), 1–5.

Wang, X., Wang, J., Wu, C., Xu, S., & Ma, W. (2022). Engineering brain: Metaverse for future engineering. *AI in Civil Engineering, 1*(1), 1–18.

Yang, Q., Zhao, Y., Huang, H., Xiong, Z., Kang, J., & Zheng, Z. (2022). Fusing blockchain and AI with Metaverse: A survey. *IEEE Open Journal of the Computer Society, 3*, 122–136.

Yin, B., Wang, Y. X., Fei, C. Y., & Jiang, K. (2022). Metaverse as a possible tool for reshaping schema modes in treating personality disorders. *Frontiers in Psychology, 13*.

Zvarikova, K., Cug, J., & Hamilton, S. (2022). Virtual human resource management in the Metaverse: Immersive work environments, data visualization tools and algorithms, and behavioral analytics. *Psychosociological Issues in Human Resource Management, 10*(1), 7–20.

REFERENCES

101 Blockchain. (2022). *10 best metaverse platforms that you can try in 2022.* https://101blockchains.com/best-metaverse-platforms/. Accessed 21 Nov 2022.

ABC News. (2022). *How the metaverse could impact the world and the future of technology.* https://abcnews.go.com/Technology/metaverse-impact-world-future-technology/story?id=82519587. Accessed 6 Nov 2022.

Abrol, A. (2022). *Metaverse vs. Multiverse—What's the difference?* https://www.blockchain-council.org/metaverse/metaverse-vs-multiverse/. Accessed 8 Oct 2022.

Accenture. (2021). *25 cloud trends for 2021 and beyond.* https://www.accenture.com/nl-en/blogs/insights/cloud-trends. Accessed 22 Nov 2022.

Accenture. (2022a). *Government enters the metaverse.* https://www.accenture.com/content/dam/accenture/final/industry/public-service/document/Accenture-Federal-Technology-Vision-2022-Government-Enters-the-Metaverse New.pdf#zoom=40. Accessed 19 Dec 2022.

Accenture. (2022b). *Protecting and serving in the metaverse continuum.* https://www.accenture.com/us-en/blogs/voices-public-service/public-safety-tech-vision. Accessed 20 Dec 2022.

Accenture. (2022c). *The next world after this: Aerospace and defence enters the metaverse.* https://www.accenture.com/_acnmedia/PDF-178/Accenture-Aerospace-Defense-Enters-Metaverse.pdf. Accessed 20 Dec 2022.

Accenture. (2022d). *Want to demystify the metaverse hype? Think of it as an internet evolution.* https://www.accenture.com/us-en/blogs/accenture-research/want-to-demystify-the-metaverse-hype-think-of-it-as-an-internet-evolution. Accessed 17 Oct 2022.

Accenture. (2022e). *Why the metaverse is a big gamechanger for defence.* https://www.accenture.com/us-en/blogs/voices-public-service/why-the-metaverse-is-a-big-gamechanger-for-defence. Accessed 20 Dec 2022.

Accenture. (2022f). *Meet me in the metaverse.* TechVision. https://www.accenture.com/_acnmedia/Thought-Leadership-Assets/PDF-5/Accenture-Meet-Me-in-the-Metaverse-Full-Report.pdf. Accessed 8 Oct 2022.

Accenture. (2022g). *Metaverse continuum set to redefine how the world operates.* https://www.accenture.com/us-en/blogs/intelligent-operations-blog/metaverse-continuum-set-to-redefine-how-the-world-operates. Accessed 21 Oct 2022.

Accenture. (2022h). *Meet me in the metaverse.* https://www.accenture.com/_acnmedia/Thought-Leadership-Assets/PDF-5/Accenture-Meet-Me-in-the-Metaverse-Full-Report.pdf. Accessed 21 Nov 2022.

Accenture. (2022i). *Why the metaverse (really) matters for travel.* https://www.accenture.com/us-en/blogs/compass-travel-blog/metaverse-travel. Accessed 21 Nov 2022.

Adgully. (2022). *The evolving face of Social Media—From socialising to the Metaverse.* https://www.adgully.com/the-evolving-face-of-social-media-from-socialising-to-the-metaverse-119631.html. Accessed 22 Nov 2022.

Afshar. V. (2022). *80% of organizations will have hyperautomation on their technology roadmap by 2024.* Digital Transformation. https://www.zdnet.com/article/80-of-organizations-will-have-hyperautomation-on-their-technology-roadmap-by-2024/. Accessed 22 Oct 2022.

Ahn, S. J., Kim, J., & Kim, J. (2022). The bifold triadic relationships framework: A theoretical primer for advertising research in the metaverse. *Journal of Advertising, 51*(5), 592–607.

Algazo, F. A., Ibrahim, S., & Yusoff, W. S. (2021). Digital governance emergence and importance. *Management, 6*(24), 18–26.

Allam, Z., Sharifi, A., Bibri, S. E., Jones, D. S., & Krogstie, J. (2022). The metaverse as a virtual form of smart cities: Opportunities and challenges for environmental, economic, and social sustainability in urban futures. *Smart Cities, 5*(3), 771–801.

Alpala, L. O., Quiroga-Parra, D. J., Torres, J. C., & Peluffo-Ordóñez, D. H. (2022). Smart factory using virtual reality and online multi-user: Towards a metaverse for experimental frameworks. *Applied Sciences, 12*(12), 6258.

Analysis Group. (2022). *The potential global economic impact of the metaverse.* https://www.analysisgroup.com/globalassets/insights/publishing/2022-the-potential-global-economic-impact-of-the-metaverse.pdf. Accessed 8 Nov 2022.

Analytics Insight. (2022a). *Welcome to the new world of art and culture with metaverse.* https://www.analyticsinsight.net/welcome-to-the-new-world-of-art-and-culture-with-metaverse/. Accessed 20 Dec 2022.

Analytics Insights. (2022b). *The metaverse—Bold plans for 2022 with Axie Infinity (AXS), and SeeSaw Protocol (SSW)*. https://www.analyticsinsight. net/the-metaverse-bold-plans-for-2022-with-axie-infinity-axs-and-seesaw-pro tocol-ssw/. Accessed 20 Nov 2022.

Anderson, J., & Rainie, L. (2022). *The metaverse in 2040*. Pew Research Centre.

Andreula, N., & Petruzzelli, S. (2022). Meta-soft power: Flipping the scales between art & culture. *RAISINA FILES*, 144.

Anshari, M., Syafrudin, M., Fitriyani, N. L., & Razzaq, A. (2022). Ethical responsibility and sustainability (ERS) development in a metaverse business model. *Sustainability, 14*(23), 15805.

Appinventiv. (2022). *How could metaverse be a game changer for the virtual gaming industry?* https://appinventiv.com/blog/metaverse-gaming/. Accessed 22 Nov 2022.

Arora, S. (2022). How the metaverse accelerates economic development for emerging economies. *Economic Times*. https://economictimes.indiatimes. com/markets/cryptocurrency/how-the-metaverse-accelerates-economic-dev elopment-for-emerging-economies/articleshow/93375549.cms?utm_source= contentofinterest&utm_medium=text&utm_campaign=cppst. Accessed 8 Oct 2022.

Automatic Sync. (2022). *What the metaverse means for cultural institutions & their digital presence*. https://www.automaticsync.com/metaverse-means-cul tural-institutions-digital-presence/. Accessed 20 Dec 2022.

Bae, J. (2021). Introduction of project-based advanced convergence structure education using metaverse in the era of future education. *Journal of Korean Association for Spatial Structures, 21*(4), 4–9.

Baker-Brunnbauer, J. (2022). *Ethical challenges for the metaverse development*.

Bala, M., & Verma, D. (2018). Governance to good governance through e-Governance: A critical review of concept, model, initiatives & challenges in India. *International Journal of Management, IT and Engineering, 8*(10), 244–269.

Bale, A. S., Ghorpade, N., Hashim, M. F., Vaishnav, J., & Almaspoor, Z. (2022). A comprehensive study on metaverse and its impacts on humans. *Advances in Human-Computer Interaction*.

Bandara. I. (2016). *The evolving challenges of internet of everything: enhancing student performance and employability in higher education*. ResearchGate. https://www.researchgate.net/figure/Four-pillar-network-connection-of-Int ernet-of-Everything-IoE-in-higher-education_fig2_299848797. Accessed 18 Oct 2022.

Barrera, K. G., & Shah, D. (2023). Marketing in the metaverse: Conceptual understanding, framework, and research agenda. *Journal of Business Research, 155*, 113420.

Bennett, D. (2022). Remote workforce, virtual team tasks, and employee engagement tools in a real-time interoperable decentralized metaverse. *Psychosociological Issues in Human Resource Management, 10*(1), 78–91.

Bernard Marr & Co. (2022). *The future of social media in the metaverse.* https://bernardmarr.com/the-future-of-social-media-in-the-metaverse/. Accessed 22 Nov 2022.

Bibri, S. E. (2022). The social shaping of the metaverse as an alternative to the imaginaries of data-driven smart cities: A study in science, technology, and society. *Smart Cities, 5*(3), 832–874.

Bibri, S. E., & Allam, Z. (2022). The metaverse as a virtual form of data-driven smart cities: The ethics of the hyper-connectivity, datafication, algorithmization, and platformization of urban society. *Computational Urban Science, 2*(1), 1–22.

Bibri, S. E., Allam, Z., & Krogstie, J. (2022). The metaverse as a virtual form of data-driven smart urbanism: Platformization and its underlying processes, institutional dimensions, and disruptive impacts. *Computational Urban Science, 2*(1), 1–22.

Binance. (2022). *Shemaroo launches "Shemaroo Theater" on decentraland.* https://www.binance.com/en/news/flash/7223148. Accessed 6 Nov 2022.

Blockchain Council. (2022a). *Decentraland metaverse: A complete guide.* https://www.blockchain-council.org/metaverse/decentraland-metaverse/. Accessed 22 Nov 2022.

Blockchain Council. (2022b). *Sandbox metaverse: An ultimate guide.* https://www.blockchain-council.org/metaverse/sandbox-metaverse/. Accessed 21 Nov 2022.

Blockchain Council. (2022c). *Understanding the seven layers of the metaverse technology.* https://www.blockchain-council.org/metaverse/seven-layers-of-the-metaverse-technology/. Accessed 6 Nov 2022.

Blockchain Works. (2021). *Axie infinity developer scores $152M in series B funding, nearing $3B valuation.* https://blockworks.co/news/axie-infinity-developer-scores-152m-in-series-b-funding-nearing-3b-valuation. Accessed 21 Nov 2022.

Blockworks. (2022). *Metaverse platforms set the record straight about daily active users.* https://blockworks.co/news/metaverse-platforms-set-the-record-straight-about-daily-active-users. Accessed 22 Nov 2022.

Bloomberg. (2021). *Metaverse may be $800 billion market, next tech platform.* https://www.bloomberg.com/professional/blog/metaverse-may-be-800-billion-market-next-tech-platform/. Accessed 21 Nov 2022.

Bloomberg. (2022). *Apple shows AR/VR headset to board in sign of progress on key project.* https://www.bloomberg.com/news/articles/2022-05-19/apple-shows-headset-to-board-in-sign-it-s-reached-advanced-stage?leadSource=uverify%20wall. Accessed 20 Nov 2022.

Bojic, L. (2022). Metaverse through the prism of power and addiction: What will happen when the virtual world becomes more attractive than reality? *European Journal of Futures Research, 10*(1), 1–24.

Bowen, J. P., & Giannini, T. (2022). *Digital experience in art and identity: The metaverse calls.*

Brabus, P (2022). *India's first metaverse marriage scheduled on February 6th in TardiWorld.* News. https://thevrsoldier.com/first-metaverse-marriage-schedu led-in-tardiworld/. Accessed 17 Oct 2022.

Buana, I. M. W. (2023). Metaverse: Threat or opportunity for our social world? In understanding metaverse on sociological context. *Journal of Metaverse, 3*(1), 28–33.

Buck, L., & McDonnell, R. (2022). *Security and privacy in the metaverse: The threat of the digital human.*

Business Insider. (2022). *AmEx reveals its metaverse ambitions in a trademark filing for tech to let people to use its payment cards in virtual worlds.* https://www.businessinsider.in/cryptocurrency/news/amex-reveals-its-metaverse-ambitions-in-a-patent-filing-for-tech-to-let-people-to-use-its-payment-cards-in-virtual-worlds/articleshow/90236616.cms. Accessed 22 Nov 2022.

Business to Community. (2022). *Disney to adopt a metaverse platform to enhance the future of storytelling.* https://www.business2community.com/nft-news/disney-to-adopt-metaverse-in-enhancing-the-future-of-storytelling-02548198. Accessed 22 Nov 2022.

Business Today. (2022). *India's first 'Metaversity' seems to be imploding from within; here's what's going on.* https://www.businesstoday.in/technology/news/story/indias-first-metaversity-seems-to-be-imploding-from-within-heres-whats-going-on-334891-2022-05-24. Accessed 22 Nov 2022.

Cai, Y., Llorca, J., Tulino, A. M., & Molisch, A. F. (2022). Compute-and data-intensive networks: The key to the metaverse. *arXiv preprint*, arXiv:2204. 02001

Cao, L. (2022). Decentralized AI: Edge intelligence and smart blockchain, metaverse, web3, and desci. *IEEE Intelligent Systems, 37*(3), 6–19.

Carter, D. (2022). Immersive employee experiences in the metaverse: Virtual work environments, augmented analytics tools, and sensory and tracking technologies. *Psychosociological Issues in Human Resource Management, 10*(1), 35–49.

Chauhan, V., Dumka, N., Hannah, E., Ahmed, T., & Kotwal, A. (2022). Recent initiatives for transforming healthcare in India: A political economy of health framework analysis. *Journal of Global Health Economics and Policy, 2*, e2022002.

Chengoden, R., Victor, N., Huynh-The, T., Yenduri, G., Jhaveri, R. H., Alazab, M., ..., Gadekallu, T. R. (2022). Metaverse for healthcare: A survey on potential applications, challenges and future directions. *arXiv preprint*, arXiv:2209.04160

Choi, H. Y. (2022). Working in the metaverse: Does telework in a metaverse office have the potential to reduce population pressure in megacities? Evidence from young adults in Seoul, South Korea. *Sustainability, 14*(6), 3629.

Christensen, L., & Robinson. A. (2022). *The potential global economic impact of the metaverse*. Analysis Group. https://www.analysisgroup.com/globalass ets/insights/publishing/2022-the-potential-global-economic-impact-of-the-metaverse.pdf. Accessed 12 Oct 2022.

CII Knowledge Summit. (2022). *Leveraging the metaverse in knowledge management*. https://ciiknowledgesummit.in/wp-content/uploads/2022/04/Met averse-for-Knowledge-Management-Paper.pdf. Accessed 20 Dec 2022.

CNBC. (2022). *Employers see promise in a metaverse workplace: Employees are a little more sceptical*. https://www.cnbc.com/2022/08/09/employers-see-promise-in-metaverse-workplace-employees-are-skeptical.html. Accessed 20 Dec 2022.

Coin Desk. (2021). *How Axie infinity creates work in the metaverse*. https://www.coindesk.com/markets/2021/07/17/how-axie-infinity-creates-work-in-the-metaverse/. Accessed 21 Nov 2022.

Coin Telegraph. (2021). *Microsoft metaverse vs. Facebook metaverse: What's the difference?* https://cointelegraph.com/metaverse-for-beginners/micros oft-metaverse-vs-facebook-metaverse-what-is-the-difference. Accessed 22 Nov 2022.

Coin Telegraph. (2022). *The feds are coming for the metaverse, from Axie Infinity to Bored Apes*. https://cointelegraph.com/news/the-feds-are-coming-for-the-metaverse-from-axie-infinity-to-bored-apes. Accessed 20 Nov 2022.

Collins, C. (2008). Looking to the future: Higher education in the Metaverse. *Educause Review, 43*(5), 50–52.

Consensys. (2021). *What is a DAO and how do they work?* https://consen sys.net/blog/blockchain-explained/what-is-a-dao-and-how-do-they-work/. Accessed 6 Nov 2022.

Damar, M. (2021). Metaverse shape of your life for future: A bibliometric snapshot. *Journal of Metaverse, 1*(1), 1–8.

Damar, M. (2022). What the literature on medicine, nursing, public health, midwifery, and dentistry reveals: An overview of the rapidly approaching metaverse. *Journal of Metaverse, 2*(2), 62–70.

Danylec, A., Shahabadkar, K., Dia, H., & Kulkarni, A. (2022). Cognitive implementation of metaverse embedded learning and training framework for drivers in rolling stock. *Machines, 10*(10), 926.

Darbinyan. (2022). *Virtual shopping in the metaverse: What is it and how will AI make it work.* Innovation. Forbes. https://www.forbes.com/sites/forbes techcouncil/2022/03/16/virtual-shopping-in-the-metaverse-what-is-it-and-how-will-ai-make-it-work/?sh=47e1a7065f27. Accessed 15 Oct 2022.

Decentraland. (2022). *Decentraland public launch.* https://decentraland.org/blog/announcements/decentraland-announces-publich-launch/. Accessed 22 Nov 2022.

Deloitte. (2021a). *The journey to government's digital transformation.* https://www2.deloitte.com/content/dam/insights/us/articles/digital-transform ation-in-government/DUP_1081_Journey-to-govt-digital-future_MASTER.pdf Accessed 21 Oct 2022.

Deloitte. (2021b). The metaverse overview: Vision, technology, and tactics. https://www2.deloitte.com/content/dam/Deloitte/cn/Documents/techno logy-media-telecommunications/deloitte-cn-tmt-metaverse-report-en-220 304.pdf. Accessed 22 Nov 2022.

Devden. (2022). Enhance warfare and training. https://www.devdensolutions. com/aerospace-defence/. Accessed 20 Dec 2022.

Dick, E. (2021). *Public policy for the metaverse: Key takeaways from the 2021 AR/VR policy conference.* Information Technology and Innovation Foundation.

Digite. (2021). Remote work in the metaverse—How will it change the way we work? https://www.digite.com/blog/metaverse-remote-work/. Accessed 22 Nov 2022.

Dionisio, J. D. N., & Gilbert, R. (2013). 3D virtual worlds and the metaverse: Current status and future possibilities. *ACM Computing Surveys (CSUR), 45*(3), 1–38.

Douglas, M. R. (2019). The string theory landscape. *Universe, 5*(7), 176.

Duan, H., Li, J., Fan, S., Lin, Z., Wu, X., & Cai, W. (2021). Metaverse for social good: A university campus prototype. In *Proceedings of the 29th ACM International Conference on Multimedia* (pp. 153–161).

Dunnett, K., Pal, S., Jadidi, Z., & Jurdak, R. (2022). The role of cyber threat intelligence sharing in the metaverse. *IEEE Internet of Things Magazine, 6*(1), 154–160.

Dwivedi, Y. K., Hughes, L., Baabdullah, A. M., Ribeiro-Navarrete, S., Gian-nakis, M., Al-Debei, M. M., ..., Wamba, S. F. (2022a). Metaverse beyond the hype: Multidisciplinary perspectives on emerging challenges, opportuni-ties, and agenda for research, practice and policy. *International Journal of Information Management, 66*, 102542.

Dwivedi, Y. K., Hughes, L., Wang, Y., Alalwan, A. A., Ahn, S. J., Balakrishnan, J., ..., Wirtz, J. (2022b). Metaverse marketing: How the metaverse will shape the future of consumer research and practice. *Psychology & Marketing, 40*(4), 750–776.

Economic Times. (2022a). *Now Shemaroo offers movie theatre on metaverse.* https://economictimes.indiatimes.com/tech/technology/now-shemaroo-offers-movie-theatre-on-metaverse/articleshow/94538342.cms?from=mdr. Accessed 6 Nov 2022.

Economic Times. (2022b). *Rise of metverse in global Pharma industry.* https://health.economictimes.indiatimes.com/news/pharma/rise-of-metaverse-in-glo bal-pharma-industry/95250523. Accessed 22 Nov 2022.

Egliston, B., & Carter, M. (2022). 'The metaverse and how we'll build it': The political economy of Meta's Reality Labs. *New Media & Society,* https://doi.org/10.1177/14614448221119785

Emergen Research. (2022). *Metaverse market, by component, by platform, by offering, by technology, by application, by end-use and by region forecast to 2030.* https://www.emergenresearch.com/industry-report/metaverse-market. Accessed 6 Nov 2022.

End Points News. (2022). *The medical metaverse is already here, but what does that mean for pharma?* https://endpts.com/the-medical-metaverse-is-alr eady-here-but-what-does-that-mean-for-pharma/. Accessed 21 Nov 2022.

Entrepreneur. (2022). *Introducing 'touch' in the metaverse.* https://www.entrep reneur.com/en-in/technology/introducing-touch-in-the-metaverse/430606. Accessed 6 Nov 2022.

Ernst & Young. (2022). *How meeting customers in the metaverse can unlock lasting value.* https://www.ey.com/en_gl/consumer-products-retail/meet-customers-in-the-metaverse-to-unlock-lasting-value. Accessed 4 Jan 2022.

European Gaming. (2021). *Search data reveals absolutely no one understands the metaverse.* https://europeangaming.eu/portal/latest-news/2021/11/02/103027/search-data-reveals-absolutely-no-one-understands-the-metaverse/. Accessed 6 Nov 2022.

Evoluteiq. (2022). *Digital banking enabled by hyperautomation and the metaverse.* https://evoluteiq.com/digital-banking-enabled-by-hyperautomat ion-and-the-metaverse/

Express VPN. (2022). *Survey reveals surveillance fears over the metaverse workplace.* https://www.expressvpn.com/blog/survey-reveals-surveilla nce-fears-over-the-metaverse-workplace/. Accessed 21 Nov 2022.

EY. (2022). *Insights on the metaverse and the future of gaming.* https://www.ey.com/en_us/tmt/what-s-possible-for-the-gaming-industry-in-the-next-dimens ion/chapter-3-insights-on-the-metaverse-and-the-future-of-gaming. Accessed 17 Oct 2022.

Facebook. (2021). Introducing horizon workrooms: Remote collaboration reimagined. https://about.fb.com/news/2021/08/introducing-horizon-workrooms-remote-collaboration-reimagined/. Accessed 22 Nov 2022.

Falchuk, B., Loeb, S., & Neff, R. (2018). The social metaverse: Battle for privacy. *IEEE Technology and Society Magazine, 37*(2), 52–61.

Fang, Z., Cai, L., & Wang, G. (2021). MetaHuman creator the starting point of the metaverse. In *2021 International Symposium on Computer Technology and Information Science (ISCTIS)* (pp. 154–157). IEEE.

Far, S. B., & Rad, A. I. (2022). Applying digital twins in metaverse: User interface, security and privacy challenges. *Journal of Metaverse, 2*(1), 8–16.

Faraboschi, P., Frachtenberg, E., Laplante, P., Milojicic, D., & Saracco, R. (2022). Virtual worlds (metaverse): From skepticism, to fear, to immersive opportunities. *Computer, 55*(10), 100–106.

Fintech Magazine. (2022). *JP Morgan is first leading bank to launch in the metaverse.* https://fintechmagazine.com/banking/jp-morgan-becomes-the-first-bank-to-launch-in-the-metaverse. Accessed 21 Nov 2022.

Forbes. (2022a). The challenges and opportunities with the metaverse. https://www.forbes.com/sites/forbestechcouncil/2022/05/17/the-challenges-and-opportunities-with-the-metaverse/?sh=38834232495f. Accessed 3 Jan 2022.

Forbes. (2022b). *The future of social media in the metaverse.* Enterprise Tech. https://www.forbes.com/sites/bernardmarr/2022/08/24/the-future-of-social-media-in-the-metaverse/?sh=3e4e24011023. Accessed 6 Oct 2022.

Forbes Digital Assets. (2022). *The world of metaverse entertainment: Concerts, theme parks, and movies.* https://www.forbes.com/sites/bernardmarr/2022/07/27/the-world-of-metaverse-entertainment-concerts-theme-parks-and-movies/?sh=498e20176531. Accessed 21 Nov 2022.

Forbes India. (2022). *What will learning in the metaverse look like?* https://www.forbesindia.com/article/take-one-big-story-of-the-day/what-will-learning-in-the-metaverse-look-like/77285/1. Accessed 22 Nov 2022.

Forbes. (2022a). 6 top metaverse coins. https://www.forbes.com/advisor/investing/cryptocurrency/top-metaverse-coins/. Accessed 20 Nov 2022.

Forbes. (2022b). A short history of the metaverse. https://www.forbes.com/sites/bernardmarr/2022/03/21/a-short-history-of-the-metaverse/?sh=6e64a9ad5968. Accessed 6 Nov 2022.

Forbes. (2022c). *Disney: The metaverse, digital transformation, and the future of storytelling.* https://www.forbes.com/sites/bernardmarr/2022/10/07/disney-the-metaverse-digital-transformation-and-the-future-of-storytelling/?sh=19849efc13c0. Accessed 21 Nov 2021.

Forbes. (2022d). *Meta's VR vs. Apple's AR strategy—Who will ultimately win?* https://www.forbes.com/sites/timbajarin/2022/10/11/metas-vr-vs-apples-ar-strategy-who-will-ultimately-win/?sh=7ae676bb44ed. Accessed 21 Nov 2022.

Gadekallu, T. R., Huynh-The, T., Wang, W., Yenduri, G., Ranaweera, P., Pham, Q. V., …, Liyanage, M. (2022). Blockchain for the metaverse: A review. *arXiv preprint*, arXiv:2203.09738

Gao, Y., Lu, Y., & Zhu, X. (2022). Metaverse, the future materials science computation platform based on metaverse. *The Journal of Physical Chemistry Letters, 14,* 148–157.

Gartner. (2017). *5 levels of digital government maturity.* https://www.gartner.com/smarterwithgartner/5-levels-of-digital-government-maturity. Accessed 19 Dec 2022.

Gartner. (2022a). *Gartner predicts 25% of people will spend at least one hour per day in the metaverse by 2026.* Press Release. https://www.gartner.com/en/newsroom/press-releases/2022-02-07-gartner-predicts-25-percent-of-people-will-spend-at-least-one-hour-per-day-in-the-metaverse-by-2026. Accessed 6 Oct 2022.

Gartner. (2022b). *What is a metaverse? And should you be buying in?* Information Technology. https://www.gartner.com/en/articles/what-is-a-metaverse. Accessed 17 Oct 2022.

Gartner. (2022c). *Metaverse evolution will be phased; here's what it means for tech product strategy.* https://www.gartner.com/en/articles/metaverse-evolution-will-be-phased-here-s-what-it-means-for-tech-product-strategy. Accessed 6 Oct 2022.

Gartner. (2022d). *Gartner predicts 25% of people will spend at least one hour per day in the metaverse by 2026.* https://www.gartner.com/en/newsroom/press-releases/2022-02-07-gartner-predicts-25-percent-of-people-will-spend-at-least-one-hour-per-day-in-the-metaverse-by-2026. Accessed 4 Jan 2022.

Gartner. (2022e). *Gartner predicts 90% of current enterprise blockchain platform implementations will require replacement by 2021.* Newsroom. https://www.gartner.com/en/newsroom/press-releases/2019-07-03-gartner-predicts-90--of-current-enterprise-blockchain. Accessed 8 Oct 2022.

Gartner. (2022f). *What is new in the 2022 Gartner hype cycle for emerging technologies.* https://www.gartner.co.uk/en/articles/what-s-new-in-the-2022-gartner-hype-cycle-for-emerging-technologies. Accessed 3 Jan 2022.

Gizmodo. (2022). *Employees are much more concerned about working in the 'metaverse' than their boss is.* https://gizmodo.com/metaverse-vr-meta-work-from-home-remote-work-1849342330. Accessed 21 Nov 2022.

Global Data (2022). *Global: Top metaverse patents holders in the aerospace and defence sector.* https://www.globaldata.com/data-insights/aerospace-and-defence/global-top-metaverse-patents-holders-in-the-aerospace-and-defence-sector-2132319/. Accessed 20 Dec 2022.

Globe Newswire. (2022). *Cryptovoxels is rebranding to voxels on May 3, 2022.* https://www.globenewswire.com/news-release/2022/05/03/2434939/0/en/Cryptovoxels-Is-Rebranding-to-Voxels-on-May-3-2022.html. Accessed 21 Nov 2022.

Goldberg, Y. (2016). A primer on neural network models for natural language processing. *Journal of Artificial Intelligence Research, 57,* 345–420.

Golf-Papez, M., Heller, J., Hilken, T., Chylinski, M., de Ruyter, K., Keeling, D. I., & Mahr, D. (2022). Embracing falsity through the metaverse: The case of synthetic customer experiences. *Business Horizons, 65*(6), 739–749.

Gorichanaz, T. (2022). Being at home in the metaverse? Prospectus for a social imaginary. *AI and Ethics*, 1–12.

Grand View Research (2018). Digital Twin Market Size, Share & Trends Analysis Report By End Use (Manufacturing, Agriculture), By Solution (Component, Process, System), By Region (North America, APAC), And Segment Forecasts, 2022 – 2030. https://www.grandviewresearch.com/industry-analysis/digital-twin-market. Accessed 10 Oct 2022.

Gray Scale (2021). The Metaverse. https://grayscale.com/wp-content/uploads/2021/11/Grayscale_Metaverse_Report_Nov2021.pdf

Grayscale Research (2022). THE METAVERSE. https://grayscale.com/wp-content/uploads/2021/11/Grayscale_Metaverse_Report_Nov2021.pdf. Accessed 18 Oct 2022.

Grigalashvili, V. (2022). E-government and E-governance: Various or Multifarious Concepts. *International Journal of Scientific and Management Research, 5*(1).

Han, D. I. D., Bergs, Y., & Moorhouse, N. (2022). Virtual reality consumer experience escapes: Preparing for the metaverse. *Virtual Reality, 26*(4), 1443–1458.

Harvard Business Review. (2022a). *How augmented reality can—And Can't—Help your brand.* https://hbr.org/2022/03/how-augmented-reality-can-and-cant-help-your-brand. Accessed 6 Nov 2022.

Harvard Business Review. (2022b). *How the metaverse could change work.* https://hbr.org/2022/04/how-the-metaverse-could-change-work. Accessed 20 Dec 2022.

Hassani, H. (2022). A study of new emerged challenges for states to rule the cyberspace: Consequences of platformization and emergence of metaverse. *Political Science, 25*(98), 161–184.

Hawkins, M. (2022a). Metaverse live shopping analytics: Retail data measurement tools, computer vision and deep learning algorithms, and decision intelligence and modeling. *Journal of Self-Governance & Management Economics, 10*(2).

Hawkins, M. (2022b). Virtual employee training and skill development, workplace technologies, and deep learning computer vision algorithms in the immersive metaverse environment. *Psychosociological Issues in Human Resource Management, 10*(1), 106–120.

Hollensen, S., Kotler, P., & Opresnik, M. O. (2022). Metaverse—The new marketing universe. *Journal of Business Strategy, 44*(3), 119–125.

Hotaran, I., Poleac, D., & Vrana, N. (2022). Five steps for sustainable business modelling in the metaverse. *Fostering Recovery Through Metaverse Business Modelling*, 530.

HR TechX. (2022). *Future is here: How metaverse becomes the part of HR technology.* https://www.hrtechx.com/2022/02/15/future-is-here-how-metaverse-becomes-the-part-of-hr-technology/. Accessed 21 Nov 2022.

Hung, H. T. B. (2022). Keep your eyes on China's metaverse: Another tool for maintaining its national security. *The Journal of Intelligence, Conflict, and Warfare*, 5(2), 1–31.

Hutson, J. (2022). Social virtual reality: Neurodivergence and inclusivity in the metaverse. *Societies*, 12(4), 102.

Huynh-The, T., Pham, Q. V., Pham, X. Q., Nguyen, T. T., Han, Z., & Kim, D. S. (2022). Artificial intelligence for the metaverse: A survey. *arXiv preprint*, arXiv:2202.10336

IBM. (2022). *Blockchain success starts here.* https://www.ibm.com/in-en/topics/what-is-blockchain. Accessed 12 Oct 2022.

Illuvium. (2022). *Home.* https://illuvium.io. Accessed 22 Nov 2022.

Inc 42. (2022). *How metaverse is reinventing healthtech & its future.* https://inc42.com/resources/how-metaverse-is-reinventing-healthtech-its-future/. Accessed 22 Nov 2022.

Inc. (2022). *According to Apple CEO Tim Cook, the next internet revolution is not the metaverse: It's this.* https://www.inc.com/nick-hobson/apple-ceo-tim-cook-next-internet-revolution-this-1-thing-metaverse.html. Accessed 21 Nov 2022.

Information Technology and Innovation Foundation. (2021). *Public policy for the metaverse: Key takeaways from the 2021 AR/VR Policy Conference.* https://itif.org/publications/2021/11/15/public-policy-metaverse-key-takeaways-2021-arvr-policy-conference/. Accessed 20 Dec 2022.

Innovius Research. (2022). *All about metaverse: 7 layers.* https://www.innoviusresearch.com/blog/market-report/7-layers-of-metaverse/. Accessed 6 Nov 2022.

Morgan, J. P. (2022). *Opportunities in the metaverse.* Content. https://www.jpmorgan.com/content/dam/jpm/treasury-services/documents/opportunities-in-the-metaverse.pdf. Accessed 12 Oct 2022.

Jena, S., Epari, V., & Sahoo, K. C. (2022). Integration of national cancer registry program with Ayushman Bharat Digital Mission in India: A necessity or an option. *Public Health in Practice*, 3, 100263.

Jiang, Y., Kang, J., Niyato, D., Ge, X., Xiong, Z., Miao, C., & Shen, X. (2022). Reliable distributed computing for metaverse: A hierarchical game-theoretic approach. *IEEE Transactions on Vehicular Technology*, 72(1), 1084–1100.

Joy, A., Zhu, Y., Peña, C., & Brouard, M. (2022). Digital future of luxury brands: Metaverse, digital fashion, and non-fungible tokens. *Strategic Change, 31*(3), 337–343.

JP Morgan. (2022). *Opportunities in the metaverse*. https://www.jpmorgan.com/content/dam/jpm/treasury-services/documents/opportunities-in-the-metaverse.pdf. Accessed on 8 Nov 2022.

Jung, Y. (2022). Current use cases, benefits and challenges of NFTs in the museum sector: Toward common pool model of NFT sharing for educational purposes. *Museum Management and Curatorship, 38*(4), 451–467.

Jungherr, A., & Schlarb, D. B. (2022). The extended reach of game engine companies: How companies like epic games and Unity technologies provide platforms for extended reality applications and the metaverse. *Social Media + Society, 8*(2). https://doi.org/10.1177/20563051221107641

Kanematsu, H., Kobayashi, T., Barry, D. M., Fukumura, Y., Dharmawansa, A., & Ogawa, N. (2014). Virtual STEM class for nuclear safety education in metaverse. *Procedia Computer Science, 35*, 1255–1261.

Karanfiloğlu, M., & Sağlam, M. (2022). Media literacy, fact-checking and cyberbullying. *Organized by, 37.*

Kim, S. H., & Yoo, J. Y. (2021). A study on the recognition and acceptance of metaverse in the entertainment industry. *Journal of the Korea Entertainment Industry Association (JKEIA), 15*(7), 1.

Kogure, J., Kamakura, K., Shima, T., & Kubo, T. (2017). Blockchain technology for next generation ICT. *Fujitsu Scientific & Technical Journal, 53*(5), 56–61.

Kshetri, N. (2022). Policy, ethical, social, and environmental considerations of Web3 and the metaverse. *IT Professional, 24*(3), 4–8.

Lee, B. K. (2021). The metaverse world and our future. *Review of Korea Contents Association, 19*(1), 13–17.

Lee, H. J., & Gu, H. H. (2022). Empirical research on the metaverse user experience of digital natives. *Sustainability, 14*(22), 14747.

Lee, J. (2022). A study on the intention and experience of using the metaverse. *Jahr: Europski časopis za bioetiku, 13*(1), 177–192.

Lee, J. Y. (2021b). A study on metaverse hype for sustainable growth. *International Journal of Advanced Smart Convergence, 10*(3), 72–80.

Lee, L. H., Braud, T., Zhou, P., Wang, L., Xu, D., Lin, Z., ..., Hui, P. (2021). From internet and extended reality to metaverse: Technology survey, ecosystem, and future directions. *arXiv e-prints*, arXiv:2110

Lee, L. H., Braud, T., Zhou, P., Wang, L., Xu, D., Lin, Z., ..., Hui, P. (2021). All one needs to know about metaverse: A complete survey on technological singularity, virtual ecosystem, and research agenda. *arXiv preprint*, arXiv: 2110.05352

Lee, L. H., Lin, Z., Hu, R., Gong, Z., Kumar, A., Li, T., …, Hui, P. (2021). When creators meet the metaverse: A survey on computational arts. *arXiv preprint*, arXiv:2111.13486

Leeway Hertz (2022a). *Digital twin and metaverse*. https://www.leewayhertz. com/digital-twin-and-metaverse/. Accessed 8 Oct 2022.

Leeway Hertz. (2022b). *Metaverse: Uplifting the virtual gaming*. https://www. leewayhertz.com/gaming-in-metaverse/. Accessed 22 Nov 2022.

Li, H., Cui, C., & Jiang, S. (2022). Strategy for improving the football teaching quality by AI and metaverse-empowered in mobile internet environment. *Wireless Networks*, 1–10.

Lim. (2022). *70% of virtual store visitors made a purchase, new study reveals*. The Industry Fashion. https://www.theindustry.fashion/70-of-virtual-store-visitors-made-a-purchase-new-study-reveals/. Accessed 15 Oct 2022.

Live Mint. (2022). *India's leading online school: 21k school is delivering the future of education*. https://www.livemint.com/brand-stories/indias-leading-online-school-21k-school-is-delivering-the-future-of-education-11654524293651. html. Accessed 21 Nov 2022.

López-Belmonte, J., Pozo-Sánchez, S., Lampropoulos, G., & Moreno-Guerrero, A. J. (2022). Design and validation of a questionnaire for the evaluation of educational experiences in the metaverse in Spanish students (METAEDU). *Heliyon, 8*(11), e11364.

Lv, Z., Shang, W. L., & Guizani, M. (2022). Impact of digital twins and metaverse on cities: History, current situation, and application perspectives. *Applied Sciences, 12*(24), 12820.

Madou, M. (2022). Now is the time to strengthen cyber defences. *Network Security, 2022*(8).

Majerová, J., & Pera, A. (2022). Haptic and biometric sensor technologies, spatio-temporal fusion algorithms, and virtual navigation tools in the decentralized and interconnected metaverse. *Review of Contemporary Philosophy, 21*, 105–121.

Marchand, A., & Hennig-Thurau, T. (2013). Value creation in the video game industry: Industry economics, consumer benefits, and research opportunities. *Journal of Interactive Marketing, 27*(3), 141–157.

Market Place Fairness. (2021). *How to invest in the metaverse in 2022*. https:// www.marketplacefairness.org/cryptocurrency/how-to-invest-in-metaverse-2022/. Accessed 6 Nov 2022.

McAllister, K. (2022). *What's the biggest effect the metaverse will have on IoT, or vice versa?* Protocol. https://www.protocol.com/braintrust/metaverse-effects-internet-of-things?rebelltitem=20#rebelltitem20. Accessed 22 Oct 2022.

Mckinsey & Company. (2022). *Value creation in the metaverse*. https://
www.mckinsey.com/~/media/mckinsey/business%20functions/marketing%
20and%20sales/our%20insights/value%20creation%20in%20the%20meta
verse/Value-creation-in-the-metaverse.pdf. Accessed 8 Oct 2022.

McKinsey. (2022). *Building a safer metaverse*. https://www.mckinsey.com/
capabilities/growth-marketing-and-sales/our-insights/building-a-safer-met
averse. Accessed 21 Nov 2022.

Medium. (2021). *The metaverse value-chain*. https://medium.com/building-
the-metaverse/the-metaverse-value-chain-afcf9e09e3a7. Accessed 6 Nov
2022.

Medium. (2021). *Cryptovoxels: Do not miss the future of the metaverse*. https://
medium.datadriveninvestor.com/cryptovoxels-this-is-the-future-of-the-met
averse-4467326d4102. Accessed 20 Nov 2022.

Meta Cat. (2022). *Metaverse analytics*. https://www.metacat.world/en-US/ana
lytics?typoe=cryptovoxels. Accessed 21 Nov 2022.

Meta Madrill. (2022). *Apple metaverse strategy; Apple's strategy for the digital
universe*. https://metamandrill.com/apple-metaverse-strategy/. Accessed 21
Nov 2022.

Metahero. (2022). *Metahero*. https://metahero.io/uploads/Metahero_WP_v3_
4.pdf. Accessed 21 Nov 2021.

Mircică, N. (2022). Immersive and engaging digital content, data visualization tools, and location analytics in a decentralized metaverse. *Linguistic &
Philosophical Investigations, 21*, 89–104.

Mohanty, L., & Swain, S. C. (2022). Use of digital technologies by the Msmes to
preserve cultural heritage of India and achieve sustainable development goals.
ECS Transactions, 107(1), 14343.

Mondaq. (2022). *India: Metaverse: Legality & regulatory concerns in India*.
https://www.mondaq.com/india/fin-tech/1195182/metaverse-legality-reg
ulatory-concerns-in-india. Accessed 19 Dec 2022.

Mourtzis, D., Panopoulos, N., Angelopoulos, J., Wang, B., & Wang, L. (2022).
Human centric platforms for personalized value creation in metaverse. *Journal
of Manufacturing Systems, 65*, 653–659.

Nalbant, K. G., & Uyanik, Ş. (2021). Computer vision in the metaverse. *Journal
of Metaverse, 1*(1), 9–12.

Nasdaq. (2022). *Disney is diving into web 3.0 and the metaverse: Here's what that
means*. https://www.nasdaq.com/articles/disney-is-diving-into-web-3.0-and-
the-metaverse%3A-heres-what-that-means. Accessed 21 Nov 2022.

Nevelsteen, K. J. (2018). Virtual world, defined from a technological perspective and applied to video games, mixed reality, and the metaverse. *Computer
Animation and Virtual Worlds, 29*(1), e1752.

News 18. (2021). *What is metaverse and why Facebook/meta thinks it's the future of internet*. https://www.news18.com/news/tech/explained-what-is-metave rse-and-why-facebook-meta-thinks-its-the-future-of-internet-4416881.html. Accessed 21 Nov 2022.

Newzoo. (2020). *The world's 2.7 billion gamers will spend $159.3 billion on games in 2020; the market will surpass $200 billion by 2023*. https://newzoo.com/insights/articles/newzoo-games-market-num bers-revenues-and-audience-2020-2023. Accessed 21 Nov 2022.

NFTically. (2022). *How will metaverse impact the global economy*. https:// www.nftically.com/blog/how-will-metaverse-impact-the-global-economy/. Accessed 18 Oct 2022.

Ning, H., Wang, H., Lin, Y., Wang, W., Dhelim, S., Farha, F., ..., Danesh-mand, M. (2021). A survey on metaverse: The state-of-the-art, technologies, applications, and challenges. *arXiv preprint*, arXiv:2111.09673

Niti Ayog. (2022). *Meta-Governance: Role of Metaverse in India's e-Governance*. https://www.niti.gov.in/index.php/meta-governance-role-metaverse-indias-e-governance. Accessed 19 Dec 2022.

Niu, X., & Feng, W. (2022). Immersive entertainment environments—From theme parks to metaverse. In *International Conference on Human-Computer Interaction* (pp. 392–403). Springer.

Norwegian Cruise Line Holdings Limited. (2022). *Norwegian cruise line announces cruise industry's first NFT collection*. https://www.nclhltd.com/ news-media/press-releases/detail/480/norwegian-cruise-line-announces-cru ise-industrys-first-nft. Accessed 6 Nov 2022.

Oh, M. J., & Kim, J. (2021). An essay on the future of metaverse as the harmony space both of Homo Ludens and Homo Fabre. *Journal of the Korea Convergence Society, 12*(9), 127–136.

Oleksy, T., Wnuk, A., & Piskorska, M. (2022). Migration to the metaverse and its predictors: Attachment to virtual places and metaverse-related threat. *Computers in Human Behavior, 141*, 107642.

Ooi, B. C., Tan, K. L., Tung, A., Chen, G., Shou, M. Z., Xiao, X., & Zhang, M. (2022). Sense the physical, walkthrough the virtual, manage the metaverse: A data-centric perspective. *arXiv preprint*, arXiv:2206.10326

Park, S. M., & Kim, Y. G. (2022). A metaverse: Taxonomy, components, applications, and open challenges. *IEEE Access, 10*, 4209–4251.

PC Mag. (2022). *What is microsoft's metaverse strategy?* https://www.pcmag. com/news/what-is-microsofts-metaverse-strategy. Accessed 22 Nov 2022.

People Matters. (2022). *How the metaverse can reshape the future of work*. https://www.peoplematters.in/article/training-development/how-the-metaverse-can-reshape-the-future-of-work-35652. Accessed 20 Dec 2022.

Pew Research Center. (2022a). *The metaverse in 2040.* https://www.pewres earch.org/internet/2022/06/30/the-metaverse-in-2040/. Accessed 3 Jan 2022.

Pew Research Center. (2022b). *About three-in-ten U.S. adults say they are 'almost constantly' online.* https://www.pewresearch.org/fact-tank/2021/ 03/26/about-three-in-ten-u-s-adults-say-they-are-almost-constantly-online/. Accessed 6 Nov 2022.

Phakamach, P., Senarith, P., & Wachirawongpaisarn, S. (2022). The Metaverse in education: The future of immersive teaching & learning. *RICE Journal of Creative Entrepreneurship and Management, 3*(2), 75–88.

Phemex. (2022). *What is metahero: The gateway to the metaverse.* https://phe mex.com/academy/what-is-metahero-hero-coin. Accessed 21 Nov 2022.

Pozniak, H. (2022). Could engineers work in the metaverse? *Engineering & Technology, 17*(4), 1–8.

Prayitno, W. (2022). The "metaverse" symbol of civilization transfer in the middle of digital economic hegemony: Synthesis of progressive law of the Covid-19 pandemic era. *International Journal of Social Science Research, 4*(3), 14–32.

Premium. (2022). *Research: The current state and predictions for the future of blockchain in the enterprise.* https://www.techrepublic.com/resource-library/ research/research-the-current-state-and-predictions-for-the-future-of-blockc hain-in-the-enterprise/. Accessed 8 Oct 2022.

PS Newswire. (2022). *Metaverse in entertainment market to reach $221.7 billion, globally, by 2031 at 32.3% CAGR: Allied market research.* https://www. prnewswire.com/news-releases/metaverse-in-entertainment-market-to-reach-221-7-billion-globally-by-2031-at-32-3-cagr-allied-market-research-301691 463.html. Accessed 20 Dec 2022.

PwC. (2022). *Making sense of bitcoin, cryptocurrency and blockchain.* https:// www.pwc.com/us/en/industries/financial-services/fintech/bitcoin-blockc hain-cryptocurrency.html. Accessed 6 Nov 2022.

Qi, P., & Chen, Z. (2022). The origin, characteristics and prospect of metaverse. *Advances in Education, Humanities and Social Science Research, 1*(1), 315–315.

Radianti, J., Majchrzak, T. A., Fromm, J., & Wohlgenannt, I. (2020). A systematic review of immersive virtual reality applications for higher educa-tion: Design elements, lessons learned, and research agenda. *Computers & Education, 147*, 103778.

Rehm, S. V., Goel, L., & Crespi, M. (2015). The metaverse as mediator between technology, trends, and the digital transformation of society and business. *Journal for Virtual Worlds Research, 8*(2), 1–6.

ReportLinker. (2022). *Global Non-fungible Token (NFT) market 2022–2026. Advanced IT Market Trends.* https://www.reportlinker.com/p06268966/Global-Non-fungible-Token-NFT-Market.html. Accessed 17 Oct 2022.

Revfine. (2022). *How the metaverse will change the travel industry.* https://www.revfine.com/metaverse-travel/. Accessed 22 Nov 2022.

Rospigliosi, P. A. (2022a). Adopting the metaverse for learning environments means more use of deep learning artificial intelligence: This presents challenges and problems. *Interactive Learning Environments, 30*(9), 1573–1576.

Rospigliosi, P. A. (2022b). Metaverse or simulacra? Roblox, Minecraft, Meta and the turn to virtual reality for education, socialisation and work. *Interactive Learning Environments, 30*(1), 1–3.

Roundhill Investments. (2021). *Roundhill's intro to the metaverse.* https://www.roundhillinvestments.com/research/metaverse/intro-to-the-metaverse. Accessed 6 Nov 2022.

Salmasi, A. V., & Gillam, L. (2009). Machine ethics for gambling in the metaverse: An. *Journal for Virtual Worlds Research, 2*(3).

Scheiding, R. (2022). Designing the future? The metaverse, NFTs, & the future as defined by unity users. *Games and Culture, 18*(6), 804–820. https://doi.org/10.1177/15554120221139218

Schmitt, M. (2022, July 21). Metaverse: Bibliometric review, building blocks, and implications for business, government, and society. *Building Blocks, and Implications for Business, Government, and Society.*

Smaili, N., & de Rancourt-Raymond, A. (2022). Metaverse: Welcome to the new fraud marketplace. *Journal of Financial Crime* Vol. ahead-of-print No. ahead-of-print. https://doi.org/10.1108/JFC-06-2022-0124.

Stanoevska-Slabeva, K. (2022). Opportunities and challenges of metaverse for education: A literature review. *EDULEARN22 Proceedings*, 10401–10410.

Statista. (2022a). *Global digital twin market share in 2020, by industry. Hardware.* https://www.statista.com/statistics/1296192/global-digital-twin-market-share-by-industry/. Accessed 8 Oct 2022.

Statista. (2022b). *Worldwide spending on blockchain solutions from 2017 to 2024. Software.* https://www.statista.com/statistics/800426/worldwide-blockchain-solutions-spending/. Accessed 18 Oct 2022.

Su, Z., Zhang, N., Liu, D., Luan, T. H., & Shen, X. (2022). A survey on metaverse: Fundamentals, security, and privacy. *IEEE Communications Surveys & Tutorials, 25*(1), 319–352.

Tang, F., Chen, X., Zhao, M., & Kato, N. (2022). The roadmap of communication and networking in 6G for the metaverse. *IEEE Wireless Communications, 30*(4), 72–81.

Tech Story. (2022). *How to access voxels metaverse (formerly cryptovoxels)?* https://techstory.in/how-to-access-voxels-metaverse-formerly-cryptovoxels/. Accessed 21 Nov 2022.

The Business Journals. (2022). *Here's how Accenture is manifesting the metaverse*. https://www.bizjournals.com/austin/news/2022/06/20/here-s-how-accenture-is-manifesting-the-metaverse.html. Accessed 22 Nov 2022.

The Conversation. (2022). *Tourism and the metaverse: Towards a widespread use of virtual travel?* https://theconversation.com/tourism-and-the-metaverse-towards-a-widespread-use-of-virtual-travel-188858. Accessed 21 Nov 2022.

The Digital Speaker. (2022). *What are the dangers of the metaverse?* https://www.thedigitalspeaker.com/what-are-the-dangers-of-the-metaverse/. Accessed 4 Jan 2022.

The Entrepreneur. (2022). *The metaverse is the future of business: Here's how to plan for it*. https://www.entrepreneur.com/science-technology/the-metaverse-is-the-future-of-business-heres-how-to/435440. Accessed 3 Jan 2022.

The Hindu. (2021). *Virtual real estate plot sells for record $2.4 million*. https://www.thehindu.com/sci-tech/technology/internet/virtual-real-estate-plot-sells-for-record-24-million/article37656785.ece. Accessed 20 Nov 2022.

The Times of India. (2022). *How the metaverse will change digital marketing*. https://timesofindia.indiatimes.com/blogs/voices/how-the-metaverse-will-change-digital-marketing/. Accessed 21 Nov 2022.

The Verge. (2021). *Meta's sci-fi haptic glove prototype lets you feel VR objects using air pockets*. https://www.theverge.com/2021/11/16/22782860/meta-facebook-reality-labs-soft-robotics-haptic-glove-prototype. Accessed 22 Nov 2022.

Theis, T. N., & Wong, H. S. P. (2017). The end of Moore's law: A new beginning for information technology. *Computing in Science & Engineering, 19*(2), 41–50.

Thomason, J. (2021). MetaHealth-how will the metaverse change health care? *Journal of Metaverse, 1*(1), 13–16.

Thomason, J. (2022). Metaverse, token economies, and non-communicable diseases. *Global Health Journal, 6*(3), 164–167.

Time. (2021). *Why TIME is launching a new newsletter on the metaverse*. https://time.com/6118513/into-the-metaverse-time-newsletter/. Accessed 5 Oct 2022.

Times of India. (2022). *Decoding the seven layers of metaverse*. https://timesofindia.indiatimes.com/blogs/voices/decoding-the-seven-layers-of-metaverse/. Accessed 6 Nov 2022.

Tong, A. (2022). Non-fungible token, market development, trading models, and impact in China. *Asian Business Review, 12*(1), 7–16.

Trade Finance Global. (2022). *Open sesame: Trade finance in the metaverse*. https://www.tradefinanceglobal.com/posts/open-sesame-trade-finance-in-the-metaverse/. Accessed 20 Dec 2022.

Travel Dine. (2022). *Six ways metaverse will impact travel and hotels*. https://www.traveldine.com/six-ways-metaverse-will-impact-travel-and-hotels/. Accessed 20 Nov 2022.

Upadhyay, A. K., & Khandelwal, K. (2022). Metaverse: The future of immersive training. *Strategic HR Review, 21*(3), 83–86.

Verdict. (2022). *Disney patent proves it is readying itself for the metaverse.* https://www.verdict.co.uk/disney-metaverse-patent/. Accessed 22 Nov 2022.

Vidal-Tomás, D. (2022). The new crypto niche: NFTs, play-to-earn, and metaverse tokens. *Finance Research Letters*, 102742.

Vox. (2019). *How robots are transforming Amazon warehouse jobs—For better and worse.* https://www.vox.com/recode/2019/12/11/20982652/robots-amazon-warehouse-jobs-automation. https://www.roundhillinvestments.com/research/metaverse/intro-to-the-metaverse

Wang, F. Y. (2022). Metavehicles in the metaverse: Moving to a new phase for intelligent vehicles and smart mobility. *IEEE Transactions on Intelligent Vehicles, 7*(1), 1–5.

Wang, G., Badal, A., Jia, X., Maltz, J. S., Mueller, K., Myers, K. J., ..., Zeng, R. (2022). Development of metaverse for intelligent healthcare. *Nature Machine Intelligence, 4*(11), 922–929.

Wang, H., Chen, D., & Deng, Q. (2022). The formation, development and research prospect of educational metaverse. *Education Journal, 11*(5), 260–266.

Wang, M., Yu, H., Bell, Z., & Chu, X. (2022). Constructing an edu-metaverse ecosystem: A new and innovative framework. *IEEE Transactions on Learning Technologies.*

Wang, X., Wang, J., Wu, C., Xu, S., & Ma, W. (2022). Engineering brain: Metaverse for future engineering. *AI in Civil Engineering, 1*(1), 1–18.

Wang, Y., Su, Z., Zhang, N., Xing, R., Liu, D., Luan, T. H., & Shen, X. (2022). A survey on metaverse: Fundamentals, security, and privacy. *IEEE Communications Surveys & Tutorials.*

Watson. (2022). *IBM global AI adoption index 2022.* https://www.ibm.com/downloads/cas/GVAGA3JP. Accessed 8 Oct 2022.

World Economic Forum. (2018). *Machines will do more tasks than humans by 2025 but Robot revolution will still create 58 million net new jobs in next five years.* News Releases. https://www.weforum.org/press/2018/09/machines-will-do-more-tasks-than-humans-by-2025-but-robot-revolution-will-still-create-58-million-net-new-jobs-in-next-five-years/. Accessed 10 Oct 2022.

Wunderman, Thompson. (2022). *Banking in the metaverse.* https://www.wundermanthompson.com/insight/banking-in-the-metaverse. Accessed 22 Nov 2022.

XR Today. (2021a). *Accenture orders record 60,000 Oculus headsets.* https://www.xrtoday.com/virtual-reality/accenture-orders-record-60000-oculus-headsets/. Accessed 21 Nov 2022.

XR Today. (2021b). *Gaming in the metaverse: The next frontier?* https://www.xrtoday.com/virtual-reality/gaming-in-the-metaverse-the-next-frontier/amp/. Accessed 21 Nov 2022.

XR Today. (2022a). Artificial intelligence in the metaverse: Bridging the virtual and real. *Virtual Reality.* https://www.xrtoday.com/virtual-reality/artificial-intelligence-in-the-metaverse-bridging-the-virtual-and-real/. Accessed 18 Oct 2022.

XR Today. (2022b). *Meta quest pro hits store shelves.* https://www.xrtoday.com/mixed-reality/meta-quest-pro-hits-store-shelves/. Accessed 6 Oct 2022.

Xu, M., Ng, W. C., Lim, W. Y. B., Kang, J., Xiong, Z., Niyato, D., ..., Miao, C. (2022). A full dive into realizing the edge-enabled metaverse: Visions, enabling technologies, and challenges. *IEEE Communications Surveys & Tutorials, 25*(1), 656–700.

Xu, X., Zou, G., Chen, L., & Zhou, T. (2022). Metaverse space ecological scene design based on multimedia digital technology. *Mobile Information Systems, 2022.*

Yahoo Finance. (2020). *Amazon's new room decorator tool lets you design a whole room with augmented reality furniture.* https://finance.yahoo.com/news/amazon-room-decorator-tool-lets-163700132.html. Accessed 6 Nov 2022.

Yang, Q., Zhao, Y., Huang, H., Xiong, Z., Kang, J., & Zheng, Z. (2022). Fusing blockchain and AI with metaverse: A survey. *IEEE Open Journal of the Computer Society, 3,* 122–136.

Yin, B., Wang, Y. X., Fei, C. Y., & Jiang, K. (2022). Metaverse as a possible tool for reshaping schema modes in treating personality disorders. *Frontiers in Psychology, 13,* 1010971.

Youtube. (2021). *The metaverse and how we'll build it together—Connect 2021.* https://www.youtube.com/watch?v=Uvufun6xer8. Accessed 17 Oct 2022.

Zeb Pay. (2022). *Banking in metaverse—The future of banking.* https://zebpay.com/in/blog/banking-in-the-metaverse. Accessed 21 Nov 2022.

Zhang, M., Zhang, Z., Chang, Y., Aziz, E. S., Esche, S., & Chassapis, C. (2018). Recent developments in game-based virtual reality educational laboratories using the microsoft kinect. *International Journal of Emerging Technologies in Learning (iJET), 13*(1), 138–159.

Zhao, Y., Jiang, J., Chen, Y., Liu, R., Yang, Y., Xue, X., & Chen, S. (2022). Metaverse: Perspectives from graphics, interactions and visualization. *Visual Informatics, 6*(1), 56–67.

Zuckerberg, M., & King, G. (2021). *Facebook launches "horizon workrooms." Here's how it works.*

Zvarikova, K., Cug, J., & Hamilton, S. (2022). Virtual human resource management in the metaverse: Immersive work environments, data visualization tools and algorithms, and behavioral analytics. *Psychosociological Issues in Human Resource Management, 10*(1), 7–20.

Index

A
Accenture, 77
Accountability, 121
Aerospace & Defence, 98
Apple, 73
Art & Culture, 100
Artificial Intelligence (AI), 35
Axie Infinity, 70

B
Banking & Finance, 81
Blockchain, 27
Broadcasting Model, 96

C
Comparative analysis model, 97
Creator economy layer, 50
Critical flow model, 96
Cryptovoxels, 70
Cyberbullying, 120

D
Decentraland, 68

Decentralization Layer, 55
Decentralized Finance (DeFi), 130
Digital Twin, 30
Discovery layer, 48
Disney, 76

E
Education & Learning, 78
E-Government, 92
Employment, 102
Experience Layer, 47

F
Facebook, 74
Fairness, 120

G
Gaming, 78

H
Healthcare & Pharmaceuticals, 80

Human Interface Layer, 58
Human Resource Management, 82
Hyperautomation, 39

I
Illuvium, 69
Infrastructure layer, 11
Interactive services model, 96
Internet of Everything (IoE), 37

K
Knowledge Management, 106

M
Metahero, 72
Metaverse, 2
Metaverse Adoption Business
    Framework, 115
Metaverse Continuum, 13
Microsoft, 75
Mobilization and Lobbying Model, 97
Multiverse, 33

N
Non Fungible Tokens (NFTs), 5

P
Privacy, 120

Public Entertainment, 106
Public Health & Safety, 104

R
Roblox, 71
Robotic Process Automation (RPA),
    39

S
Safety, 122
Sandbox, 68
Security, 123
Social issues, 121
Social Media & Entertainment, 83
Spatial Computing Layer, 53

T
Technology Hype Cycle, 132
Trade & Economy, 101
Travel & Tourism, 81

V
Virtual reality (VR), 2, 33

W
Web 2.0, 24
Web 3.0, 24